Digital Television
DVB-T COFDM
and
ATSC 8-VSB

Second Edition

Mark Massel

First Published by digitalTVbooks.com
(1st edition published 2000, 2nd edition 2003)

ISBN number: **0 - 9704932 - 1 - 5**

Contents

II COFDM principles

III   Set Top Box (STB) architecture

V   DVB-T COFDM Vs ATSC 8-VSB

VI   HDTV, the future and related topics

1.0   Introduction

1.1   Scope of this book

There are many publications and books available that describe digital television with relation to the consumer advantages such as: More consumer choice, better quality images and sound, the simultaneous delivery of wide screen format, sophisticated electronic program guides, high definition TV. Also the promise of future services such as near video on demand, home shopping, internet access, games etc etc. It is however the aim of this book to look at the technical detail of how all these services are actually transmitted to the consumer.

After a brief historical introduction, the book is split into seven main parts: Part I concentrates on the structure of the actual programming data that must be transmitted. The structure is the *digital transport stream* and is the same format for digital terrestrial, satellite and cable alike. The fundamental theory of MPEG compression is also explained. Part II then looks in detail at the actual transmission mechanism for digital terrestrial signal transmissions as defined by the DVB-T for transmissions first adopted in Europe. A full explanation of coded orthogonal frequency division multiplexing (COFDM) is given and its DVB-T implementation, along with all the steps needed to transmit the signal. Part III goes on to describe the actual architecture (hardware and software) of the set top boxes needed to receive and decode the DVB signals. Included here is an introduction to the very complex MHP middleware. Part IV then looks at the US ATSC system for the transmission of digital TV, employing the 8-VSB technology. The DVB-T and the ATSC systems are then compared technically in part V. Part VI then describes the programming formats such as HDTV, that are, in part, major driving forces for the move to digital transmission techniques in the first place. As an aid to a deeper understanding of the digital transmission techniques of COFDM and VSB the final part, part VII gives some standard theory underlying the techniques. In particular Fourier transform theory is detailed here.

1.2 Historical background to OFDM

OFDM (orthogonal frequency division multiplex) is certainly not a new theory. It grew out of the work done by the US military on multi carrier modulation (MCM) back in the late 1950's. This work was in the area of high frequency radio. The original OFDM patent was filed in 1966 by R. W. Chang in the US (Patent number 3,488,445) and granted four years later. OFDM is in fact a special form of MCM. The basis of MCM is to split the data up into a number of parallel streams, then these streams are used to modulate a number of different carriers. The clever thing about OFDM is that it overcomes a number of potential problems associated with the use of a large number of carrier frequencies. For example the generation of these frequencies is performed in a digital manner and the reception of these frequencies, so close together, without some sort of interference would be difficult without the use of expensive, high quality band pass filters with steep edges. (This is dealt with by orthogonality, the 'O' in OFDM). OFDM solves these problems too.

1.3 Historical background to VSB

VSB (vestigial side band) is a form of amplitude modulation (AM) which was adopted as the US standard for digital television transmission in 1996. It was recommended by the FCC's advisory committee on advanced television services. The use of signals outside of the data carriers for synchronisation makes this a rugged system, allowing for a synchronised well locked picture even if the data is corrupt. A consortium known as the *grand alliance* was formed in 1993 to lobby the FCC to adopt the VSB system for the delivery of digital standard resolution television (STV) as well as digital *high definition television* (HDTV). This consortium comprised of the following groups: AT&T, David Sarnoff Research

Centre, General Instruments Corporation, Massachusetts Institute of Technology, Philips Electronics North America Corporation, Thomson Consumer Electronics, and Zenith Electronics Corporation. After running various trials they recommended the trellis coded 8-VSB system for digital terrestrial transmissions and 16-VSB for cable transmissions. This book will describe the trellis coded 8-VSB digital terrestrial system.

1.4   Technological advances making digital TV possible

There have been various advances in digital technology which have all come together that have made digital terrestrial TV, as well as digital TV in general, possible. These, simply put, are: (i) The development of the MPEG-2 (Moving Pictures Expert Group) video compression standard. This takes advantages of similarities in the picture from one frame to another, as well as compressing the actual data in the same frame. (ii) In the case of digital terrestrial COFDM, it has only recently become viable to produce semiconductor devices fast and cheap enough to perform the FFT (Fast Fourier Transform) operation that is an essential aspect of OFDM as will be seen later. (iii) Even if the technology is available this also has to be available at the right cost to consumers, to create the market. This has also only recently been possible with advances in the semiconductor manufacturing processes.

In 1992 an organisation called DVB (Digital Video Broadcasting) was set up to set standards in all areas of digital television broadcast in Europe. This voluntary group, made up of over 200 organisations, published the digital satellite (DVB-S) specification and digital cable (DVB-C) specification in 1994. The digital terrestrial (DVB-T) specification was then finalised in late 96 / early 1997.

Prior to the finalisation of the DVB-T specification much work was carried out by the European *digital terrestrial television broadcast* (dTTb) project. This project resulted in the building of a demonstrator that showed the many advantages of a COFDM based system, particularly in countering echoes and reflections. To move this research on, the *digital video broadcasting integrated receiver decoder* (DVBird) project was initiated. Its aim was to realise the first terrestrial *integrated receiver decoder* (IRD), otherwise known as a terrestrial *set top box* (STB). An important result of the DVBird work was the partitioning of the circuit functions into a four chip solution. In 1995 the *Digital Television Group* (DTG) was set up in the UK to take the DVB standard, and turn it into a working system to meet the UK government plans for digital TV deployment in the UK. This successful work culminated in the first consumer digital terrestrial receivers adhering to the DVB-T specification being in the market, and with the UK consumer in late 1998.

1.5   The DVB (Digital Video Broadcasting) group

The DVB has played a leading and extremely successful role in the proliferation of digital television around the world. It has defined the standards in all area from broadcast systems and infrastructure to specifications enabling interactivity. And more recently to providing solutions for horizontal multimedia markets.

Although this book is specifically concerned with the DVB-T specification, the reader is referred to the **DVB Specifications Overview** at the end of this book which details all the specifications created to make digital television a reality.

I Programming Data Stream Structure

2.0   MPEG 2 transport stream

The main purpose of a transport stream is (due to its short and fixed length), the ease of implementing the forward error correction (FEC) system.

The transport stream is the bit stream that carries all the programming information in all three of the digital broadcast media and will be now be described here in more detail.

The transport stream has been defined in such a way as to minimise the processing effort required at a receiver when performing the following operations:

1)   Retrieval of coded data from one program within the transport stream, decode it and present the results.
2)   Extraction of a single program from the transport stream, to produce a new stream containing only that stream.
3)   Extraction of the transport stream packets of one or more programs, from one or more transport streams and to output a different transport stream.
4)   Extraction of the contents of one program from the transport stream and to produce as output a program stream containing that one program
5)   The conversion of a program stream into a transport stream in order for transmission over a lossy environment, and then the recovery of a valid, and in certain cases, identical program stream.

All the various fields associated with the transport packet header and payload are shown here, but for further, more detailed information, the reader is recommended to read the relevant MPEG-2 systems specification. The transport stream is the digital bit stream that carries all the information that a particular service provider transmits. The transport stream, as shown in *figure 2.1*, is a multiplex of all the programs available.

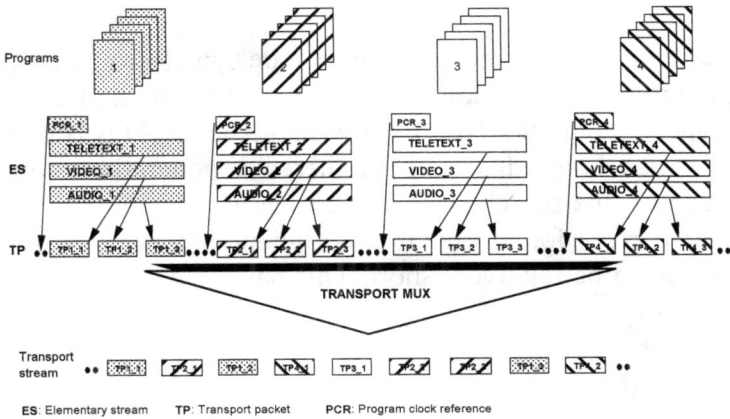

Figure 2.1

ES: Elementary stream   TP: Transport packet   PCR: Program clock reference

Each program will consist of various items; typically video, audio in various languages, and teletext. These programs are constructed out of what are known as *elementary streams* (ES's). These are the compressed data streams bundled together with a common time reference, the *program clock reference* (PCR). (It is usual to associate a PCR with each program, however it is also possible to have a PCR associated with a number of programs, and even transmitted within its own transport packet). The PCR is an important part of the demux operation and will therefore be further explained in section 9.1.7.1 of this book. In order for the ES's to be transmitted down the same channel they must be split up into small sections: These sections are called the *transport packets* (TP's). These TP's are then multiplexed together to produce one bit stream; the *transport stream* (TS). The structure of a transport packet is shown in *figure 2.2*. Note that *figure 2.1* shows the PCRs with their own TP's within in TS. This is feasible and sometimes the case, however in the majority of applications they are actually inside the video ES.

2.1   The Transport packet

The transport packet has been designed to transport programming information from a transmitter, be that on land or in space, to the receiver, usually in the home. This is termed the *channel*. Large distances are therefore generally involved. Because of this the channel is reasonably error prone. To be able to correct for errors in an efficient manner the packet length has to be relatively short. This has therefore been set to 188 bytes.

Synchronisation
byte (#47)

1	187 Bytes

Data bytes

188 byte transport packet

Figure 2.2

The transport packet is always of length 188 bytes. However the adaptation field can vary from 1 to 184 bytes if present. (Note that the packet that is physically transmitted to a receiver is 204 bytes in length due to the addition of 16 bytes of error correcting Reel Solomon bytes). The packet is split into a header and a payload. The payload, which will be described later, contains the PES and the *program specific information* (PSI). The shorter the adaptation field length, the more payload data there will be. All the fields associated with the header will be described here, but only the most important of the fields associated with the payload. In both cases much more detail is available in the following publications:

1)   ISO/IEC 13818-1. This document explains the transport packet structure
2)   ETS 300 468. This document goes into more detail regarding the program specific information (PSI).

2.2 The transport packet header

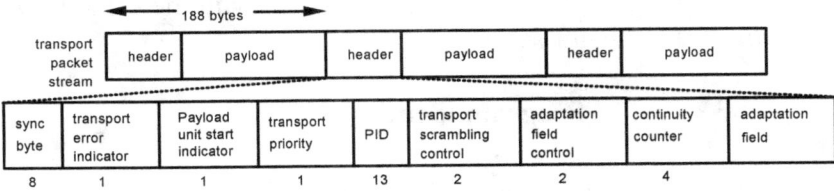

sync byte	transport error indicator	Payload unit start indicator	transport priority	PID	transport scrambling control	adaptation field control	continuity counter	adaptation field
8	1	1	1	13	2	2	4	

Figure 2.3

The header consists of the synchronisation byte, followed by other fields (generally termed the *prefix*), these will now be explained.

2.2.1 Synchronisation byte

The first byte in the transport packet is always the synchronisation byte, of value hex 47. This is used at the receiver to detect the start of the packet.

2.2.2 Transport error indicator

This single bit flag, if high, indicates that at least one uncorrectable bit error exists in the associated transport stream packet.

2.2.3 Payload unit start indicator

This single bit flag indicates the start of PES or PSI data, in packets that contain this data. If this data is contained in the packet, either the PES packet data will start immediately after this flag and header, or an optional PES header will be present followed by the PES data. See section 2.3 on the PES payload.

2.2.4 Transport priority indicator

This single bit flag, when high, indicates that the associated packet has a higher priority that packets with the same PID but with the indicator set to zero.

2.2.5 Program identifier (PID)

The most important field is the 13 bits defining the *program identifier* (PID). For every one of the compressed video, audio or teletext streams there is a unique PID value. That is to say, that each elementary stream has its own, and only one PID. Some PID values are however reserved. These are reserved for transport stream packets which carry service information (SI) and are as shown in *table 2.1*:

PID Value (hex)	Description
0	Program association table (PAT)
1	Conditional access table (CAT)
2 to F	Reserved
10	Network information table (NIT) and stuffing table (ST)
11	Service description table (SDT), bouquet information table (BAT), and stuffing table (ST)
12	Event information table (EIT), and stuffing table (ST)
13	Running status table (RST) and stuffing table (ST)
14	Time/date table (TDT), time offset table (TOT), and stuffing table (ST)
15	Network synchronisation
16 to 1B	Reserved for future use
1C	Inband signalling
1D	Measurement

1E	DIT
1F	SIT

Table 2.1

2.2.6 Transport scrambling control

This two bit field indicates the scrambling mode of the transport stream packet payload. Note that the transport stream packet header, and adaptation field (if present) cannot be scrambled.
If the value is zero, then there is no scrambling. (This is also the case if null packets are transmitted). The remaining values; 1, 2 and 3 will have meanings that are broadcaster defined. That is to say, these will have meaning to the particular conditional access system used.

2.2.7 Adaptation field control

This two bit field indicates whether the associated transport stream packet header is followed by an adaptation field and / or payload as shown in *table 2.2*:

Value (hex)	Description
0	Reserved for future use by ISO/IEC
1	No adaptation field, payload only
2	Adaptation field only, no payload
3	Adaptation field followed by payload

Table 2.2

In the case of a null packet, this field will be set to 1. Note also that the receiver will discard any packets with this field set to zero.

2.2.8   Continuity counter

This four bit field increments with each transport stream packet with the same PID value. Once the maximum value has been reached, it will wrap back around to zero. It is however not incremented if the adaptation field control of the packet is zero. Also, if a duplicate or NULL packet is sent, the continuity count is not incremented, and the adaptation field control value will have the value of either 0 or 1.

2.2.9   Adaptation field

This field, as has been said, may or may not be present and can vary from 1 to 184 bytes in length if present.

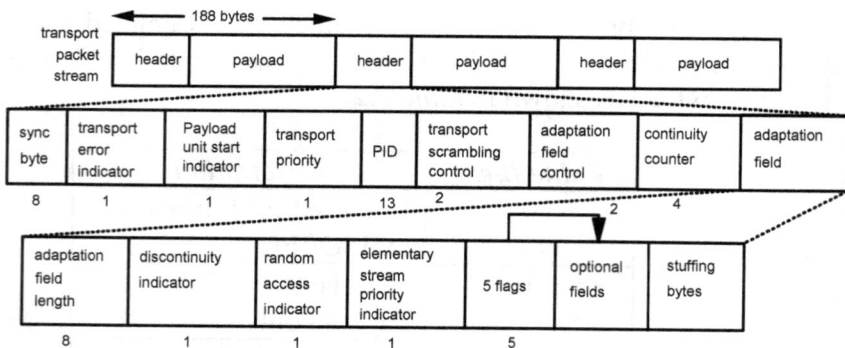

Figure 2.4

As can be seen the adaptation field contains a number of fields that will now be explained.

2.2.9.1 Adaptation field length

This 8 bit field specifies the number of bytes in the adaptation field immediately following. *Table 2.3* shows some of the values that are allowed:

Value (hex)	Description
0	Single stuffing byte inserted
2	Adaptation field length is in the range 0 to 182
3	Adaptation field length is 183

Table 2.3

2.2.9.2 Discontinuity indicator

This single bit flag indicates system time base discontinuities and continuity counter discontinuities. When high it indicates that the discontinuity state is true for the current transport stream packet. When low, or not present, then the discontinuity state is false.

2.2.9.3 Random access indicator

This single bit flag indicates that the current and subsequent transport packets with the same PID contain some information to aid random access at this point.

2.2.9.4 Elementary stream priority indicator

This single bit flag indicates the priority of the elementary stream data carrier within the payload of this transport packet. If high, this indicates that the payload is of a higher priority than the payloads of other transport stream packets. In the case of video this field will only be set high if the packet contains one or more bytes from an intra-coded slice. If this flag is low, this indicates that this payload has the same priority as all other packets which do not have this bit set high.

2.2.9.5   Adaptation field flags

These 5 flags are used to extend the length of the adaptation field to include extra information such as time stamps.

1)   If the *PCR* flag is set high, then the adaptation field contains a PCR field coded in two parts.

2)   If the *OPCR* flag is set high, this indicates that the adaptation field contains an OPCR field coded in two bytes.

3)   If the *splicing point* flag is set high, this indicates that a *splice countdown* field will be present in the associated adaptation field, specifying the occurrence of a splicing point.

4)   If the *transport private data* flag is set high, this indicates that the adaptation field contains one or more private data bytes.

5)   If the *adaptation field extension* flag is set high, this indicates the presence of an adaptation field extension.

2.2.10   Optional fields

The optional fields as indicated by the flags mentioned above, and shown in the diagram of *figure 2.5*, have the following meanings:

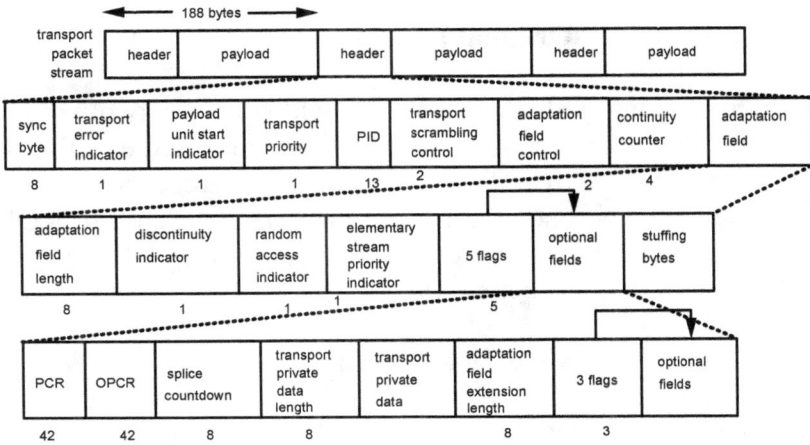

sync byte	transport error indicator	payload unit start indicator	transport priority	PID	transport scrambling control	adaptation field control	continuity counter	adaptation field
8	1	1	1	13	2	2	4	

adaptation field length	discontinuity indicator	random access indicator	elementary stream priority indicator	5 flags	optional fields	stuffing bytes
8	1	1	1	5		

PCR	OPCR	splice countdown	transport private data length	transport private data	adaptation field extension length	3 flags	optional fields
42	42	8	8	8	8	3	

Figure 2.5

2.2.10.1   Program clock reference (PCR), and extension

This is a very important field. It is of length 42 bits, coded in two parts. It indicates the intended time of arrival of the byte containing the last bit of the program clock reference base at the input of the system target decoder. The first field is of length 33 bits and is the *program clock reference base*. The second field is of 9 bits in length and is the *program clock reference extension*. The value of the PCR is the PCR base x 300 + the PCR extension.

2.2.10.2 Original program clock (OPCR)

The optional OPCR is a 42 bit field coded in two parts. The two parts, the base and the extension are coded exactly as the PCR stated above. This field allows the reconstruction of a single transport stream from another. The presence of a OPCR is indicated by the OPCR flag. OPCR's are only coded into transport stream packets that include PCR's.

2.2.10.3 Splice countdown

The splice countdown flag is of length 8 bits. It is a signed field, and when positive specifies the remaining number of transport stream packets, of the same PID, required to reach a splicing point. When negative, it indicates that the associated transport stream packet is the nth following the splicing point.

2.2.10.4 Transport private data length

This 8 bit field specifies the number of private data bytes immediately following the private transport data length field. There will never be enough of these to extend beyond the adaptation field

2.2.10.5 Private data bytes

These 8 bit fields are not specified by ISO/IEC

2.2.10.6 Adaptation field extension

This 8 bit field indicates the number of bytes of extended adaptation field data following the end of this field, including reserved bytes if present.

2.2.10.7   Optional field flags

There are three of these single bit flags indicating the following:

1)   If the *ITW* flag is set high, this indicates the presence of the *itw offset* field
2)   If the *piecewise rate* flag is set high, this indicates the presence of the *piecewise rate* field.
3)   If the *seamless splice* flag is set high, this indicates that the splice type and *DTS next AU* fields are present.

The fields associated with the above flags are shown in *figure 2.6*:

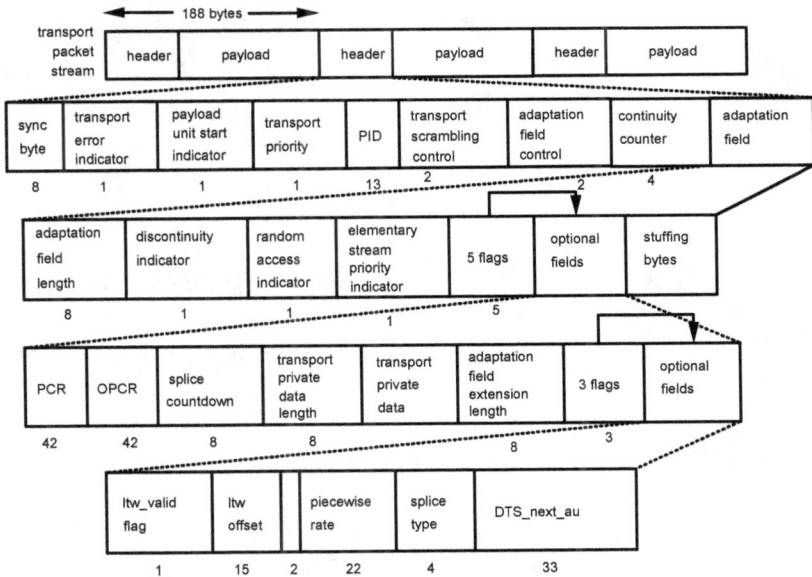

Figure 2.6

2.3   The transport packet payload

We have now seen all the fields associated with the transport packet header. Now we will look at only the PES payload fields that are of most interest. The various fields associated with the PES payload are shown in *figure 2.7*.

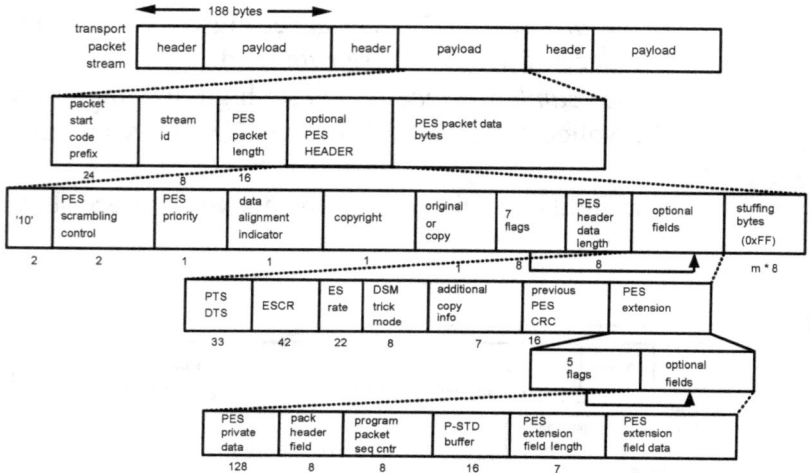

Figure 2.7

Note that the last block of fields in *figure 2.7* will not be explained in this book. The reader is advised to read the ISO/IEC specification as detailed at the top of this section for more information.

2.3.1   PES scrambling control

Either the whole transport stream can be scrambled (with the exception of the packet header and the adaptation field), or just the PES field. (Note that if PES field scrambling is done, the PES packet header, including the optional fields will not be scrambled). The PES scrambling control field is two bits in length and indicates the following: If the value is zero, then there is no scrambling. If value 1 is reserved for future DVB use. The value 2 indicates that the packet is scrambled with an even key, and 3 an odd key.

2.3.2 Presentation time stamp (PTS) and decoding time stamp (DTS) fields

These are related to decoding times. The PTS field is of length 33 bits and is encoded in three fields. This indicates the time of presentation to the decoder of an elementary stream. The DTS field is of length 33 bits and is also coded in three fields. The value indicates the decoding time of an elementary stream. The time stamps are used to correctly synchronise related ES's at the decoder.

2.4 Program specific information (PSI)

Program specific information is information that is needed by a decoder in order to allow the data to be de-multiplexed. This data will not be scrambled (With the exception of the event information table (EIT), which may be), it is broken up into sections, and transmitted in the transport stream. These sections however will not necessarily be neatly mapped onto the transport stream packets. They will, in general, cover a number of transport packets. A section has a header and a payload that will not necessarily fill the transport packet completely. Either the section header or payload may well straddle two or more transport packets. Private data can also be transmitted using exactly the same structure as the PSI data. ie transmitted as sections.

The sections when they are received at the decoder, are used to construct tables, these tables are known as *program specific information* (PSI) tables. These four main tables give information regarding the multiplex from which they are transmitted. Another six tables that may or may not be constructed also give information regarding services and events carried by other multiplexes, and even on other networks. These tables are the service information (SI) tables and will be described on completion of this section on the transport packet fields.

2.4.1 Program association section (PAS)

There are four tables defined which carry program specific information (PSI). These tables are made up of what are known as *sections*. Sections are a maximum of 1024 bytes in length, (with the exception of private sections which can be 4096 bytes). A table can be made up of a maximum of 256 sections. The repetition rate as to how often these tables are sent in the transport stream is not specified. However these must be sent fairly regularly since the receiver will need to construct these tables to know which PID's to extract based on what service or program the viewer has selected. We will first look at the various fields in the transport packet associated with recovering the sections and will then look at the contents of the tables in more detail later.

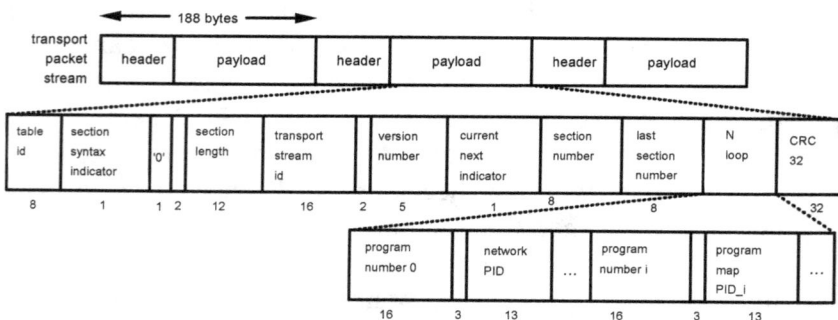

188 bytes

| transport packet stream | header | payload | header | payload | header | payload |

table id	section syntax indicator	'0'	section length	transport stream id	version number	current next indicator	section number	last section number	N loop	CRC 32
8	1	1	2 12	16	2	5 1	8	8		32

program number 0	network PID	...	program number i	program map PID_i	...
16	3 13		16	3 13	

Figure 2.8

Looking back at the PID field in the header. If this field is set to zero, then the packet payload will contain the program association table (PAT). In fact this is the very first PID that the receiver will look for. It will look at the payload of all packets with PID set to zero and extract the sections to build up the PAT. The table provides the association between a particular program number and the PID value of the transport stream packets which carry the program definition. The various fields as shown in *figure 2.8* are now looked at.

2.4.2   Table ID

This 8 bit field identifies the content of a transport stream PSI section. For a program association section it has the value zero, as shown in *table 2.4*:

Table IDValue (hex)	Description
0	Program association section
1	Conditional access section
2	TS program map Section
3	TS description section
4 to 3F	Reserved
40	Network information section - actual network
41	Network information section - other network
42	Service description section - actual transport stream
43 to 45	Reserved for future use
46	Service description section - other transport stream
47 to 49	Reserved for future use

4A	Bouquet association section
4B to 4D	Reserved for future use
4E	Event information section - actual transport stream, present / following
4F	Event information section - other transport stream, present / following
50 to 5F	Event information section - actual transport stream, schedule
60 to 6F	Event information section - other transport stream, schedule
70	Time / date section
71	Running status section
72	Stuffing section
73	Time offset section
74 to 7D	Reserved for future use
7E	Discontinuity information section
7F	Selection information section
80 to FE	Broadcaster defined
FF	Reserved

Table 2.4

2.4.3 Section syntax and length

The *single bit* syntax flag has the value '1'.

The *section length* is a 12 bit field which specifies the number of remaining bytes in the section after this field and including the CRC. The first two bits will be set to zero. The section length will not exceed 1021 bytes, so that the entire section has a maximum length of 1024 bytes.

2.4.4 Transport stream ID and version number

The *transport stream ID* is a 13 bit field and acts as a label to identify the transport stream from others within a network.

The 5 bit *version number* field, indicates the current version of the whole of the PAT. It is incremented by one whenever the version of the PAT changes, until the value 31 is reached. The counter will then wrap around back to zero.

2.4.5 Current next indicator

This single bit flag signifies the applicability of the current PAT. When high the PAT is applicable, when low it indicates that the PAT is not yet applicable, but will be the next table to become valid.

2.4.6 Section and last section number

The *section* is an 8 bit field, which gives the number of the section. The first section of the PAT will be hex 0. It will be incremented by 1 with each additional section in the PAT.

The *last section number* is an 8 bit field that specifies the number of the last section in the PAT.

2.4.7   Program number (*n*)

This 16 bit field specifies the program to which the program map PID is applicable. If set to zero the following reference will be a network PID. For all other values the field is defined by the broadcaster. The field will not take any single value more than once within one version of the PAT.

2.4.8   Network PID

This 13 bit field specifies the PID of the transport stream packets which will contain the network information table. The value of the network PID field is defined by the broadcaster, but will have certain specified values. The network PID is optional.

2.4.9   Program map PID

This 13 bit field specifies the PID of the transport steam packets that contain the program map section applicable for the program specified by the program number. No program number will have more than one program map PID assignment. The values of the program map PID's are defined by the broadcaster, but can only take on certain specific values

2.4.10   CRC 32

This 32 bit field contains the cyclic redundancy check (CRC) value that gives zero output after processing the entire program association section.

2.4.11   Program map section (PMS)

The program map section data will be used to construct the program map table (PMT) at the receiver. This will be explained in section 2.5 of this book.

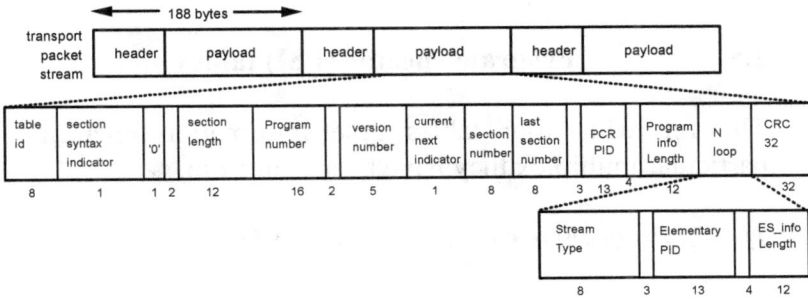

Figure 2.9

2.4.12   **Private section**

As with the other section data, these sections will be used to construct private tables at the receiver / decoder.

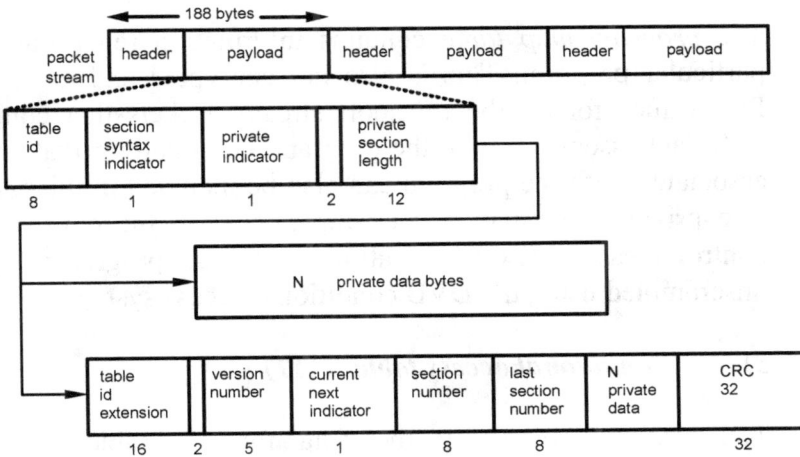

Figure 2.10

Although the diagram of *figure 2.10* shows the field structure for the transmission of private data sections, the same structure is used for the transmission of the other PSI table sections, and so will not be repeated. The various SI tables will now be looked at.

2.5   Program specific (PSI) tables

The following four tables give information regarding the particular multiplex they have been sent from.

1)   *Program association table (PAT)*

The sections containing the data for this table are not encrypted and are transmitted in transport packets with PID value hex 0. The table gives the link between the program number and the PID of the transport packet carrying the program map table (PMT).

2)   *Program map table (PMT)*

The *program map table* contains information regarding one particular program. This is also not encrypted and gives the PID values for all the transport stream packets that contain PES data associated with the program. Optionally private data associated with the program can also be included in this table. The private data may, for example, contain the entitlement control message (ECM) to allow encrypted programs to be unscrambled using the DVB conditional access system.

3)   *Conditional access table (CAT)*

The sections containing the data for this table are not encrypted and are transmitted in transport packets with PID value hex 1. This table gives the PID values of the transport packets containing the *entitlement management message* (EMM). This is the second piece of information needed by the DVB conditional access system for decryption purposes.

4)   Network information table (NIT)

This table gives information regarding a network that is made up of more than one transport stream. It gives information regarding the multiplexes and characteristics of the network.

The most important tables are therefore the PAT and the PMT. These can be better understood by referring to *figure 2.11*.

Program association table (PAT)
PID: hex 0

Program 0	NIT PID: hex 33
BBC1	PID: hex 320
BBC2	PID: hex 200
Channel 4	PID: hex 220
Channel 5	PID: hex 235
etc	etc

Program map table (PMT)
of Channel 4: PID: hex 220

PCR_PID	PID: hex 218
Video	PID: hex 110
Audio English	PID: hex 121
Audio Spanish	PID: hex 115
ECM (descr. key)	PID: hex 106
etc	etc

Figure 2.11

One of the first things a receiver / decoder must do when it powers on is to filter for all transport packets with PID value hex 0. The decoded will then know what programs are available for the viewer to access. So, for example, if the viewer wanted to view channel 4, he or she would select channel 4 via the electronic program guide (EPG). The viewer may also select to have the audio in Spanish say. The decoder would then use the PAT to see which PID's it needs to access for channel 4. The example above gives the channel 4 PMT as being transmitted in transport packets with PID value hex 220.

The decoder would then filter all the transport stream packets with PID value hex 220 and construct the PMT. It would then use this table to find the PID values of the packets containing the PCR PID (218), the video PID (110), the Spanish audio PID (115), possibly the ECM PID (106), etc. (Most decoders can filter a minimum of 32 PID values at one time). Since the advent of MHP, also included in this table are references to the PID values that contain applications to enable interactive services.

2.6   Service information (SI) tables

The following tables give information regarding services and events carried by the multiplex, and also by other multiplexes, and even on other networks.

1)   Bouquet association table (BAT)

This table groups services together that can be presented to the viewer via the EPG. These services may be from more than one multiplex. The table data is transmitted in transport stream packets with PID value hex 11, and table ID of hex 4A.

2)   Service description table (SDT)

This table lists various parameters associated with a service. Such as an indication as to the status of a particular program, ie running, paused, starting in a few minutes (ie for video recording) and also descriptors describing the service. The table data is transmitted in transport stream packets with PID value hex 11, and table ID's hex 42 and 46.

3)   Event information table (EIT)

This table gives information, in chronological order, regarding events; such as start time, and duration of the event. The table data is transmitted in transport stream packets with PID value hex 12, and table ID's of hex 4E, 4F, 50 to 5F, and 60 to 6F.

4)   Running status table (RST)

This table allows rapid updates of the timing status of one or more events. Events such as schedule changes. A separate table for this allows for a faster update. The table data is transmitted in transport stream packets with PID value hex 13, and table ID hex 71.

5)   Time and date table (TDT)

This table, which is transmitted in a single section, contains the time and date, and so allows the decoder to be updated with the real time. The table data is transmitted in transport stream packets with PID value hex 14 and table ID hex 70.

6)   Time offset table (TOT)

This table, which is transmitted in a single section, contains the time and date information and a local time offset . The table data is transmitted in transport stream packets with PID value hex 11 and table ID hex 73.

7)   Stuffing table (ST)

This table is used to invalidate section data at a system delivery boundary such as a cable head end. When one section of a sub table is overwritten, then all the sections of that sub table will also be overwritten (stuffed) in order to retain the integrity of the section number field.

All tables are transmitted with differing repetition times to allow the STB to quickly learn all about the particular services it can receive. PAT and PMT tables are usually transmitted very regularly whereas, for example, the TOT and TDT tables only relatively rarely; maybe once every minute or half minute.

3.0   MPEG 2 compression

MPEG is a very powerful compression technique for video data streams. If no compression were to be performed a data rate of around 200 Mbits/s would need to be sustained for the delivery of standard digital TV programming. This would take of the order of 5 or 6 times the bandwidth available. MPEG stands for; *Moving pictures expert group*. It was set up to standardise a coding method for the compression of moving pictures and sound. The group, set up by ISO (*International Standards Organisation*), and the IEC (*International Elecrotechnical Commission*), publish the specification under the four standards.

1)   13818-1 Systems

This defines a multiplex structure for combining audio and video data and means of representing the timing information needed to replay synchronised sequences in real time.

2)   13818-2 Video

This specifies the coded representation of video data and the decoding process required to reconstruct pictures

3)   13818-3 Audio

This specifies the coded representation of audio data. Note that there are other standards that are also generally used in the compression of audio. See section 17.0 of this book for more information on these.

4)   13818-4 Conformance

This specifies the procedures for determining the characteristics of coded bit streams and for testing compliance with the requirements stated in the above. Since there are many text books that go into great detail on the subject of

MPEG, the aim of this section is only to give the reader an understanding of the basic principles involved.

3.1   MPEG levels and profiles

The first standard, MPEG-1 was published in 1988, and was primarily for the compression of video data at bit rates of 1.5 Mbits/s. This was targeted at CD-ROMs, and high bit rate digital terrestrial land lines. This was then extended, in MPEG-2, for higher bit rates. There are a number of *profiles* and *levels* defined that allow different encoders and decoders to be built, allowing variations in bit rate, picture resolution and complexity (therefore cost) of decoding devices needed. MPEG defines tools that allow the reconstruction of the video from only some pieces of the bit stream. The bit stream is therefore constructed in layers with a base layer onto which more refinement is added in subsequent layers. A profile is a subset of the tools mentioned above. A level is a set of constraints on parameters such as picture size and bit rate.

4 levels are defined for different resolutions; these are as shown in *table 3.1*:

Level	Maximum bit rate	Max. number of samples
High	80 Mb/s	1920 x 1080 x 30 1920 x 1152 x 25
High - 1440	60 Mb/s	1440 x 1080 x 30 1440 x 1152 x 25
Main	15 Mb/s	720 x 480 x 30 720 x 576 x 25
Low	4 Mb/s	352 x 240 x 30 352 x 288 x 25

Table 3.1

To complement these resolutions and bit rates, 4 profiles are also defined, these are:

1)   Simple profile

This is designed to simplify the encoder and decoder, at the expense of higher bit rates. This method does not use B frames.

2)   Main profile

This is a compromise between compression rate and decoder complexity. It uses all three of the frame types I, P and B.

3)   SNR Scalable profile

This allows the transmission of basic quality pictures to be transmitted (with variations in quantisation) with supplementary information allowing the picture characteristics to be enhanced. Therefore allowing data to be transmitted hierarchically. This means programming can be sent either to:

    (i)   both high and standard definition receivers
    (ii)   poor quality areas, where either basic data can be received or enhanced quality where possible.

4)   Spatially scalable

This allows the transmission of basic quality pictures to be transmitted (with variations in spatial resolution) with supplementary information allowing the picture characteristics to be enhanced.

5)   High profile

This is intended for HDTV applications in 4:2:0 and 4:2:2 formats. The numbers mentioned above under the High profile description refer to the ratio of Y, C_b and C_r samples. Where Y is the luminance signal, C_b is the blue chrominance signal or colour difference, and C_r is the red chrominance or colour difference. These are related to red (R), green (G) and blue (B) by the following equations:

$$Y = 0.587 \, G + 0.299 \, R + 0.114 \, B$$
$$C_b = 0.564 \, (B - Y)$$
$$C_r = 0.713 \, (R - Y)$$

The transformations were made when colour TV was first proposed to satisfy the following two criteria:

1)   The colour signals when viewed on a black and white TV had to produce good results.

2)   The signal had to be transmitted with the same bandwidth as the existing channel.

It can therefore be seen that by selecting a particular profile at a particular level the following types of MPEG-2 encoding / decoding specifications can be defined:

HP @ HL
HP @ H14L
HP @ ML
SSP @ H14L
SNRP @ ML
SNRP @ LL
MP @ HL
MP @ H14L
MP @ ML
MP @ LL
SP @ ML

The one used for standard TV transmissions is the main profile at main level (MP @ ML). The high profile specifications are intended for HDTV. In particular MP @ HL.

3.2   The compression algorithm

Generally speaking there are two different types of compression algorithm; lossy and loss less. A good example of a loss less algorithm is the *run length encoder* (RLE). This algorithm takes, for example, a string of 1's and 0's, and instead of saving the entire number string, it reduces the number of bytes needed to store the data by simply counting how many 1's there are together in a row, them storing that number. This system works well if there are long strings of 1's or 0's as is found in simple pictures. This technique does not degrade the data being encoded. Irrespective of the number of times the algorithm is run on the resulting data. MPEG-2 is a lossy algorithm. If the same algorithm is run time and time again on the same picture or data set, the picture or data would soon degenerate, such that the resulting decoding would bear very little resemblance to the original picture or data set.

Three types of pictures are defined for the MPEG-2 algorithm. These are I frame, P frame and B frame pictures:

3.2.1   I (intra) frames

These are frames that are coded with reference to only one still picture and to no other pictures. This is similar to the JPEG algorithm for still pictures. Therefore all the information from these frames can be used at a decoder to completely reconstitute the original picture. These frames are therefore used as reference pictures at the receiver / decoder.

3.2.2   P (predicted) frames

These frames are coded from the preceding I or P frame pictures using motion compensation prediction. This will be explained later on.

3.2.3   B (bi-directionally predicted) frames

These frames are coded by directional interpolation between the I or P frame pictures which precede and follow them. These frames do not introduce errors and give high compression rates. Pictures must therefore be stored at the encoder to allow predictions to be made from future as well as past pictures, prior to transmission.

A typical transmission frame sequence that may be sent after MPEG-2 encoding is shown in *figure 3.1*:

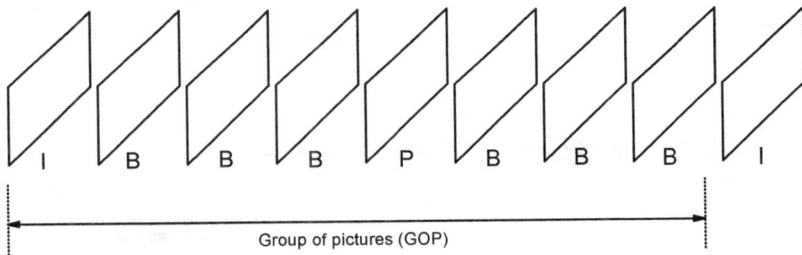

Figure 3.1

The I frame comes first, followed by 3 B frames in the example of *figure 3.1*, then another P frame, and another 3 B frames. This group of 8 frames is known as a group of pictures (GOP). Defined since the frames are sandwiched between two I frames. The decoder can be built that will decode, just I frames, I and P frames or all three frames, resulting in very different results.

3.2.4   Motion compensation

Motion compensation is an *inter-frame* method. That is to say it operates over more than one frame. The method derives the value of a particular pixel in one frame from the pixel in another previous one. The difference from this predicted value and the actual value is then transmitted. In most pictures this difference, or error value will be small and so constitutes a small amount of information to send, hence a good compression ratio is achieved.

The prediction process is improved by using a *macroblock* of pixels in the picture field (16 x 16 pixels) and compared to 16 x 16 pixels in a predefined search area in the previous field. The block that gives the 'best match' in terms of the smallest error is used. This is effective since it takes into account the movement of an object from one picture to another. This is motion compensation. (Note that this motion compensation can be done with a frame, and the previous frame, or a frame further back). The vector value that describes the 'best match' block in the predicted frame is coded and also transmitted to the decoder. The following *figure 3.2* shows the block diagram of the inter-frame predictive coding:

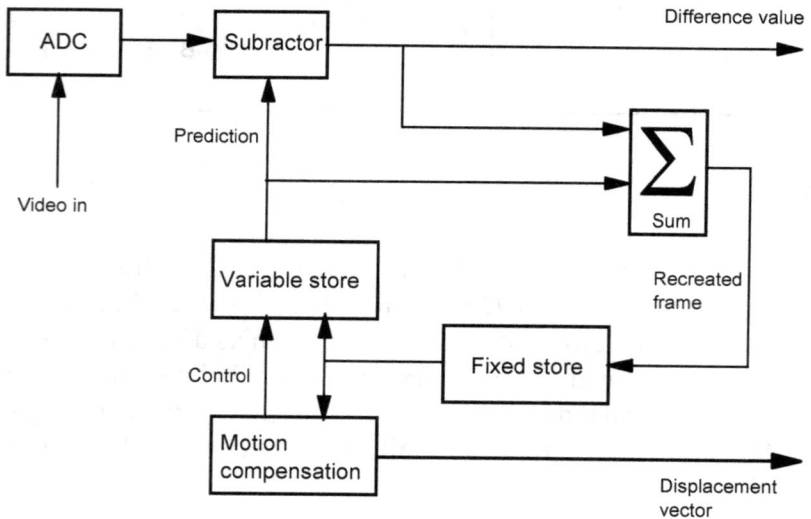

Figure 3.2

The fixed store holds the previous frame. The variable store is used for block matching. To start off with the fixed store is initially filled with null values. The current frame is then coded without reference to a predicted frame. This therefore establishes a reference for the decoder. Such an *intra-coded* frame is transmitted every so often to prevent any accumulation of prediction or transmission errors.

3.2.5 Discrete cosine transform (DCT) coding

Once the motion compensation has been achieved, the pictures are compressed (using the DCT technique). This is termed *intra-frame* coding.

The technique effectively converts (or transforms) the picture information from the spatial domain to the frequency domain. Once in the frequency domain some information can be lost due to the eye not being very sensitive to high frequency spatial components. The way it works it that the picture is split up into 8 x 8 pixel blocks. Take for example one of these blocks which has density or dark levels described by the numbers shown in *figure 3.3*:

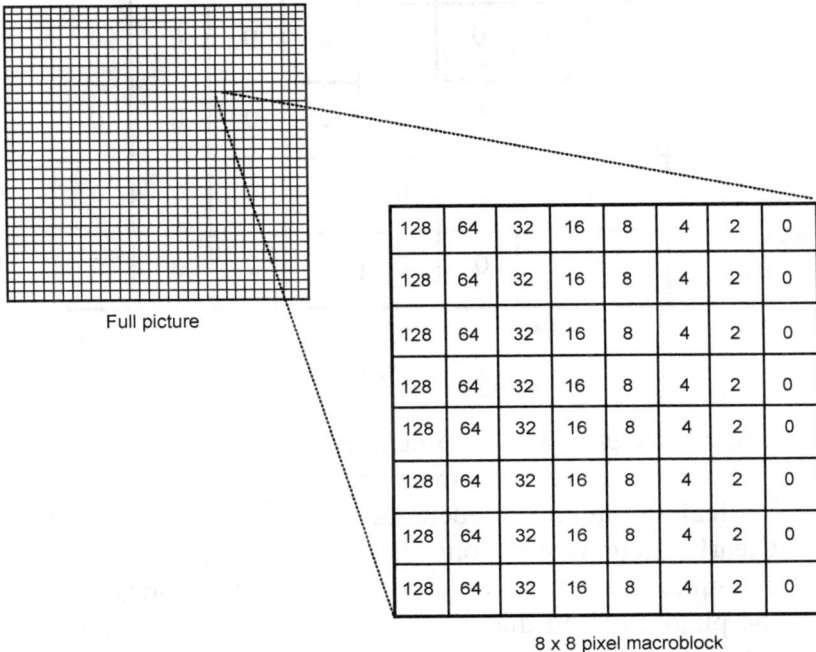

Full picture

128	64	32	16	8	4	2	0
128	64	32	16	8	4	2	0
128	64	32	16	8	4	2	0
128	64	32	16	8	4	2	0
128	64	32	16	8	4	2	0
128	64	32	16	8	4	2	0
128	64	32	16	8	4	2	0
128	64	32	16	8	4	2	0

8 x 8 pixel macroblock

Figure 3.3

As we know from Fourier analysis, it is possible to express any spatial signal in terms of its frequency components. This is achieved using the discrete Fourier transform (DCT).

Each frequency component will in general have a different amplitude. It is these amplitudes that are referred to as the coefficients, and these coefficients are the parameters that are transmitted instead of the actual spatial picture information. This may look something as shown in *figure 3.4.*

32.9	-3.2	0	-2.6	0	-1.4	0	0
0	0	0	0	0	0	0	0
0	0	0	0	0	0	0	0
0	0	0	0	0	0	0	0
0	0	0	0	0	0	0	0
0	0	0	0	0	0	0	0
0	0	0	0	0	0	0	0
0	0	0	0	0	0	0	0

Figure 3.4

The coefficients shown going horizontally represent increasing frequencies increasing from left to right, and those going vertically represent frequencies increasing from top to bottom. Clearly there is only one component in the vertical in this example, since there are no density (or amplitude) changes in the pixel block in this direction. The value of the coefficients will obviously change as the picture goes from left to right. But generally speaking they will reduce going from left to right, also going from top to bottom.

With accurate enough calculation of the coefficients no data need be lost when performing the DCT transformation. However the eye will not perceive any loss of image quality if some of the higher frequency components are lost. Therefore a particular threshold is set, below which data is thrown away. The remaining coefficients are then quanitsed.

3.2.6   Coefficient quantisation

This process is equivalent to performing an analogue to digital conversion. A particular number can be represented by a digital code with a certain number of bits. The higher the number of bits the more accurate is the representation of the original number. Clearly the more bits used the more information needs to be transmitted. Therefore fewer bits are used for the higher frequency coefficients since the eye is less sensitive to these. This quantisation is actually changed in real time to regulate the bit rate used to transmit the picture, in this way there is no possibility that more information is transmitted than there is bandwidth to accommodate. Generally the quantisation will vary depending on other factors too, such as if the block has come from an inter-frame or an intra-frame coded picture. Also the picture content plays a part. If the content is a very gradually changing tone, then this must be quantised with more accuracy since the eye would otherwise notice the difference. *Figure 3.5* shows a typical non linear quantisation curve:

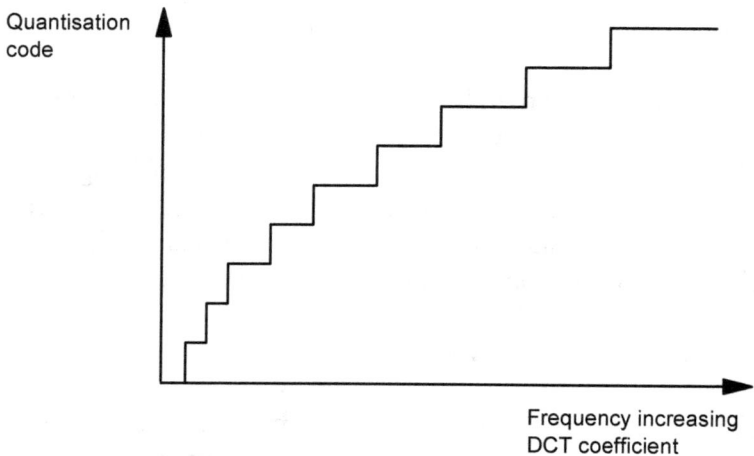

Quantisation code

Frequency increasing
DCT coefficient

Figure 3.5

Since these may vary on a block by block basis, this information must clearly also be transmitted to the receiver. The way this quantisation is performed will vary depending on the equipment used at the encoding side.

3.2.7   Run length encoding (RLE)

The next stage in the compression process is to *run length encode* the quantised data. Since the quantisation may result in many 0's due to the thresholding performed prior to the quantisation. It makes no sense to transmit all these 0's, so instead a value is transmitted to say how many 0's there are in a row. This in essence is what RLE is all about. The 'row', as mentioned above, or more accurately, stream of data, is actually taken in a zig zag fashion as shown in *figure 3.6*:

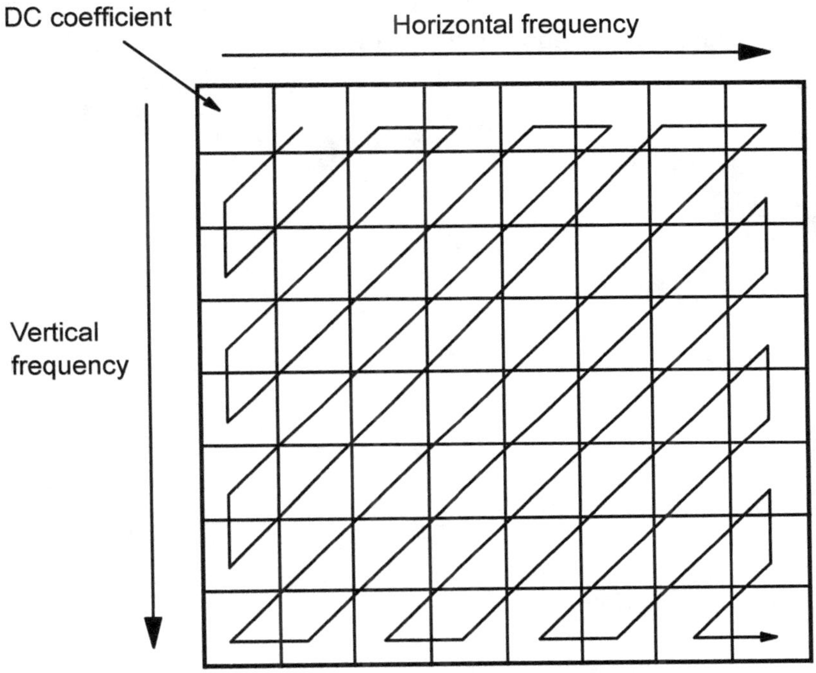

Figure 3.6

This produces a greater compression ratio since the zig zag is the most optimum smooth way of moving from the low to the high frequency part of the coefficient matrix. There is more probability of finding strings of 0's together in this way. Note that the first coefficient is the DC coefficient and is simply sent as is.

3.2.8   Variable length coding (VLC)

VLC or *Huffmann coding* is the last step in the compression process. This simply encodes the most frequently occurring values with short length binary numbers, and the less frequently occurring ones with longer ones. This clearly has a compression benefit. In fact the RLE and VLC together are responsible for compression of 2 or 3.

4.0   Error correction

All digital systems by definition consist of strings of zeros and ones. With physical hardware and non ideal transmission media the ones and zeros can often become swapped. If this happens too often then the digital system may no longer be able to perform its task adequately. A simple system could simply send the data three times, and the receiving device could implement a voting regime to reconstruct the correct data sequence. However, this technique is very wasteful in bandwidth. More powerful error correction techniques were therefore devised by people such as Shannon, Hamming, Reed, and Solomon. The basic ideas of those used in digital television will be looked at here.

Once the 188 byte packets have been produced they need to be transmitted to the receiver with as few errors as possible. The number of errors that are allowed are in the order of one error in every 10^{10} bits transmitted, or better. The *bit error ratio* (BER) is defined as the reciprocal of this, ie 10^{-10}. A communications channel with this type of BER is known as a *quasi error free* (QEF) channel.

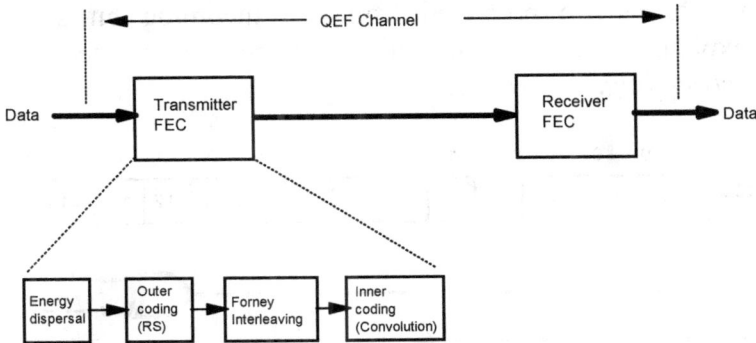

Figure 4.1

The major blocks of a digital terrestrial (and digital satellite) error correction system are shown in *figure 4.1*. The general term used to describe the error correction is *forward error correction* (FEC).

The term *forward* is used since this is a correction method which manipulates the data prior to transmission, and so prior to any errors being introduced. An example of a non-forward error correction system is a bi-directional channel where if an error is detected at the receiver, the erroneous data can be asked to be re-sent. The FEC processing, at the transmitter, allows the receiver to correct the vast majority of errors. In basic terms the data is first pseudo randomised then some redundancy is added, ie a certain level of data is copied, the data is then re-ordered so as to spread out any errors that may be clumped together. The data stream is then finally transmitted. The functional blocks in the *figure 4.1* have the following functions:

4.1   **Energy dispersal**

This is a method of removing long runs of 0's or 1's in the signal (Such as NULL packet stuffing bytes) which would otherwise give the receiver a problem at the other end of the channel. This is achieved at the bit level. The pseudo random sequences are produced from the incoming bit stream as explained below. Note this is sometimes termed as *data scrambling. Figure 4.2:*

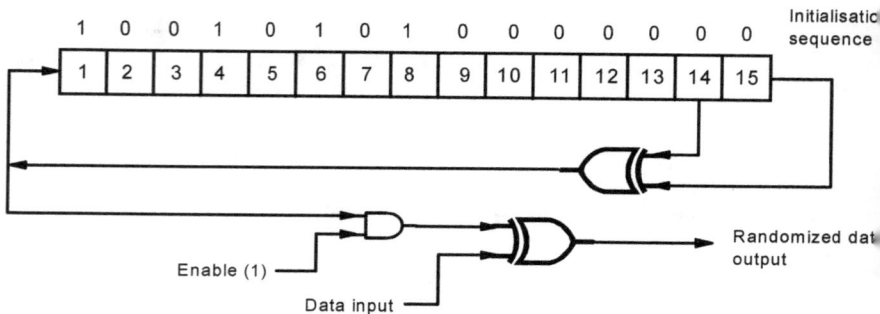

The pseudo random sequences are generated by using the following polynomial:

$$1 + X^{14} + X^{15}.$$

This is realised, as shown in *figure 4.2*, with a shift register. Bits 14 and 15 are fed into an exclusive OR gate, then back to the input and also exclusive OR'ed with the data, when enabled, to give the output. The initialisation sequence is as shown in the diagram and is loaded every 8 transport packets. Note that this gives a pseudo random binary sequence (PRBS) data output. It is important to realise that this is not truly random. Since it is known how the randomising (or scrambling) is done, this is used at the receiver side to reconstitute the original data. In fact not all the data is scrambled. The exceptions are:

1)   The first byte of an 8 transport stream sequence.

      The first byte of a transport stream is known as the synchronisation byte and is hex 47. This is inverted (and not scrambled) to give hex B8. This allows the descrambler at the receiver side to know that it is at the start of an 8 packet sequence.

2)   The first bytes of the subsequent 7 transport stream packets.

      Subsequent transport packet synchronisation bytes (after the first of a sequence of 8 packets) are left alone, ie at hex 47. This is achieved by setting the enable bit of the scrambler to zero during these byte times. This is to allow the start of a packet to be identified at the receiver.

4.2   Outer coding (Reed-Solomon coding)

Reed-Solomon coding is mathematically very complicated and so will not be explained here in any detail. The aim is simply to give an overview of the coding technique. The coding is a block level code. That is to say that it operates over a block of data. The block of data must therefore be constructed prior to the code operation being performed. This puts an overhead on the system in terms of memory and the need to block synchronise. The coding adds 16 additional bytes to the 188 byte transport stream packets, making a final transport stream packet size of 204 bytes. This error correction algorithm, as applied to a transport stream, is characterised by the three numbers n, k and l, where n is the number of bytes in the final transport stream. k is the number of bytes of the original transport stream, and t is the number of bytes that can be corrected. So:

$n = 204$ (final transport packet length)
$k = 188$ (Original transport packet length)
$t = 8$ (Number of correctable bytes)

This is sometimes shortened to RS(204,188).
Note that Reed-Solomon can be applied for different packet lengths and with different error correction bytes, for example the US ATSC system uses RS(207, 187), ie:

$n = 207$ (final transport packet length)
$k = 187$ (Original transport packet length)
$t = 10$ (Number of correctable bytes)

The selection of the DVB-T Reed-Solomon system was made based on the transport packet size. This being 188 bytes. It was decided to use the same transport stream format and so the same packet size as that for the DVB-S system (digital satellite).

Reed-Solomon are actually special cases of the more general *Bose, Chaudhuri, Hocgenghem* (BCH) codes.

The first digital broadcast system for satellite data transmission was in fact the US DSS format. This system uses shorter transport packets and an equivalent RS / FEC . The DVB-T RS(204, 188) is actually a shortened version of the RS(255, 239). 51 Null packets are added onto the 188 byte transport stream packets to produce 239 byte blocks. These are then sent throughout the RS(255, 239) encoder adding 16 parity bytes. Hence 255 protected byte blocks are produced. After the encoding the 51 null bytes are discarded giving the 204 byte protected packets.

The Reed-Solomon code effectively specifies a polynomial by generating a large number of points. The Reed-Solomon code detects errors within these points much as the human eye can detect errors in what should be a smooth curve. The Reed-Solomon code can detect and correct up to $(n-k)/2$ errors. The field generator polynomial used is:

$$P(X) = X^8 + X^4 + X^3 + X^2 + 1$$

with the code generator polynomial:

$$G(X) = (X + \alpha^0)(X + \alpha^1) \ldots \ldots (X + \alpha^{15})$$

These are the same as is specified for digital satellite (DVB-S).

4.3　　　　　Forney interleaving

Outer interleaving or *Forney convolution interleaving* is performed to basically spread out the errors and so make the outer coding more effective. As stated above, the outer coding can correct up to 8 bytes in a transport stream packet. Clearly if a burst error condition occurs, ie a burst of energy from some noise source, then more than 8 bytes within the same packet could become corrupted. The Forney convolution effectively takes these errors and spreads them out (in time) over a number of packets (generally only two), thus allowing the outer coding to be more effective.

Figure 4.3 shows the transmitter architecture. Data is input from the Reed-Solomon outer coding and output to the convolution inner coder.

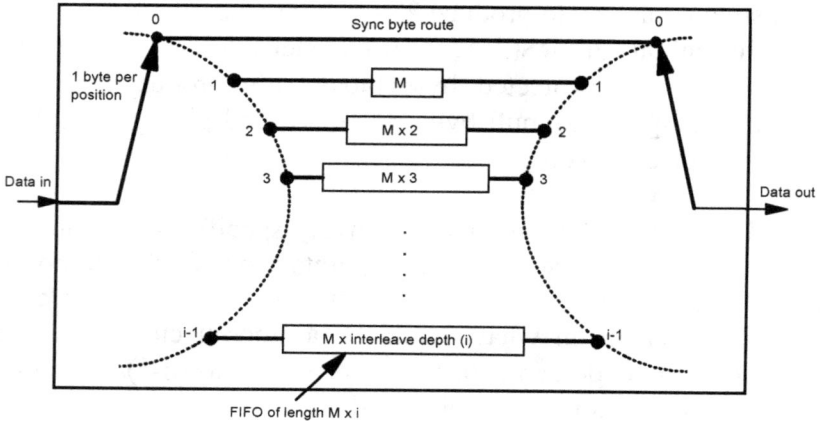

Figure 4.3

The following are the dimensions of the interleaver for the DVB-T system:

i = 12 (Number of branches, therefore shift registers)
L = 204 (Length of the packet to be protected)
M = (L/i) x j (Size of the FIFO's in bytes)
j (An index that ranges from 0 to i-1)

Therefore there are 12 individual branches, with the largest FIFO being of length 187 bytes. Input bytes will therefore be delayed by 17, 34, 51, ... 187 bytes, depending on the byte index. Note that the sync byte always passes through directly and therefore experiences no delay. *Figure 4.4* shows the architecture at the receiver side.

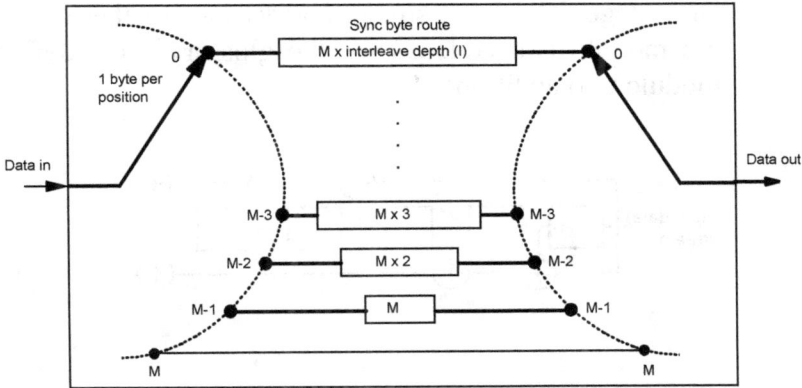

Figure 4.4

Data bytes with j index of 11 were delayed by 187 bytes at the transmitter, and so at the receiver the j = 11 bytes experience no delay, while the j = 0 bytes experience the maximum delay of 187 bytes. The original data stream is therefore reconstituted.

4.4   Inner coding (convolution)

This code is sometimes called *Viterbi coding* after the author of the decoding algorithm generally used. Convolution codes operate at the bit level rather than block level such as Reed-Solomon codes. This has the advantage of the generator (or decoder) not having to store a whole block of data in expensive memory prior to performing the coding (or decoding). The operation is *on the fly*. (Some memory is however required, this along with computational power rises exponentially with efficiency). The operation of the coding is as follows: *n* output data streams are produced from the input stream. Generally *n* is set to 2 or 3. (Such systems are also described as 1/2 or 1/3 rate) These streams are not merely direct copies of the input steam. This could be done but would produce a very inefficient system. The input stream is fed into shift register stages which have intermediate output taps after each stage. The input stream and various of these output taps are modulo two added. (See the glossary for a definition of modulo two addition).

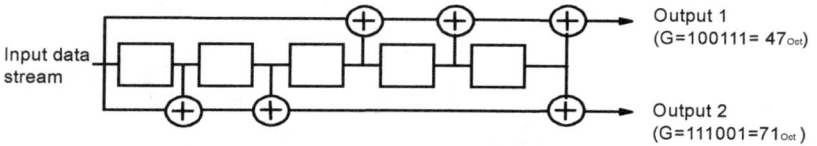

Figure 4.5

The modulo two adders are inserted at different positions within the shift register block. *Figure 4.5* shows an example with $n = 2$. The polynomials in the example of *figure 4.5* are:

$$G_1 = 1 + X^3 + X^4 + X^5$$
$$G_2 = 1 + X + X^2 + X^5$$

By combining the polynomial coefficients in groups of three the octal format is generated. It is the octal format that is used to describe the particular polynomial. In this example:

$$G_1 = 47 \text{ Oct}$$
$$G_2 = 71 \text{ Oct}$$

Each polynomial has different error correcting properties and so the selection of the best ones is important to increase the probability of reconstructing the correct sequence.

The DVB-T system actually uses 6 shift register stages and the generator polynomials:

$$G_1 = 171 \text{ Oct}$$
$$G_2 = 133 \text{ Oct}$$

The architecture as shown in *figure 4.5* can be considered as being a state machine. With each bit that is input 2 bits are output. Since there are 6 stages in the DVB-T implementation, this gives rise to 64 states. It is the change from one state to another, that, when drawn as a diagram gives rise to the term *trellis* diagram. The operation of which is explained in more detail in section 14.4.5 of this book on trellis coding.

4.4.1   DVB-T error probability

Section 14.4.5.1 describes the operation of the convolution encoder in detail for the ATSC 8-VSB implementation. This uses the Hamming distance concept to calculate the distance apart of groups of bits. This is a measure of the probability of error. Since QAM style modulation is used for the DVB-T system, the distance d_s is used instead. This can be used in conjunction with the description of section 14.4.5.1 to gain a better understanding of the operation of the DVB-T convolution encoder operation.

When the amplitude and the phase are changed, as in any QAM system, this is done to allow the receiver to more easily identify individual constellation points. The distance between constellation points can be considered as being related to the probability of the introduction of errors into the system. For example *figure 14.7* shows a simple QPSK signal constellation. The distance d_s is a measure of the probability of errors being introduced. Therefore if this distance can be increased then errors will be less likely to occur. This is the purpose of the QAM modulation technique. However when an error does occur, it is more probable that this error will be small rather than large. Ie a small distance from the ideal constellation point position.

4.4.2 Puncturing

The coding talked about in the last section is used after symbol mapping (2 bits per symbol in the case of QPSK). Puncturing is the act of removing some bits from the group or G outputs in *figure 4.5*. The puncturing block shown in *figure 4.6* doesn't always select all of the simultaneous bits from the G_1 and G_2 outputs, but drops some to a greater or lesser extent. This is defined as being the *puncture rate*.

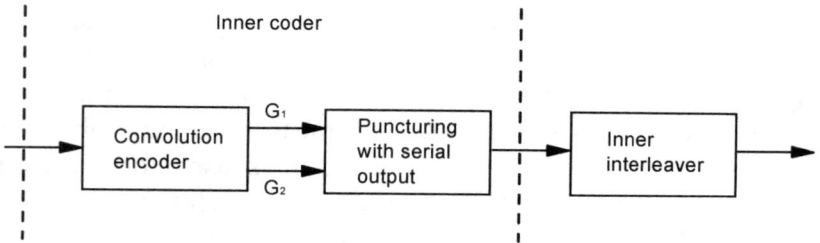

Figure 4.6

This obviously degrades the redundancy, but greatly increases the useful data transmission rate. Taking the example used previously with $n = 2$. The useful data rate with no puncturing would be 1/2 of the original rate (prior to convolution encoding). If the puncturing removes every 4th bit, (this is known as a puncture rate of 3/4) the data rate then becomes (1/2)/(4/3), or 2/3 of the original rate. Code rates of 1/2, 2/3, 3/4, 5/6 and 7/8 are allowed for DVB-T.

II   COFDM Principles

5.0 COFDM description

What is COFDM ? It stands for Coded Orthogonal Frequency Division Multiplexing. The coded part refers to the error correcting techniques described in part I of this book. OFDM is the method by which the data is actually transmitted over the airwaves, so lets first look at the frequency and Electromagnetic (EM) spectrum.

5.1 Frequency spectrum

The frequency spectrum is shown in *figure 5.1*:

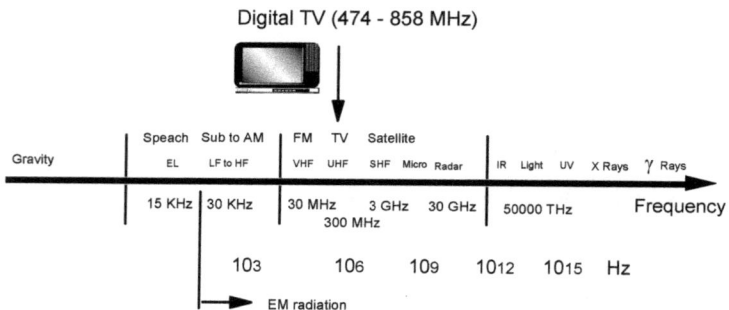

Figure 5.1

As can be seen, the frequency spectrum goes all the way from the almost undetectable gravity waves at one end to the high energy gamma rays and beyond at the other. Within this vast spectrum, in which certain phenomena can be associated with wave properties, there lies the electromagnetic (EM) waves. These are, at the basic level, produced whenever points of charge (like electrons) change their velocity. The speed at which these waves travel is exactly the same, and is c, or the speed of light, about 3×10^8 meters per second (in a vacuum like space). The EM spectrum starts with radio waves at around 30 Khz, and goes up to higher frequency and so higher energy gamma rays. As the frequency gets higher the properties of the waves change. For example radio and TV frequencies can easily pass through structures like buildings, whereas infra red and light can't. Radio waves are propagated everywhere, microwaves are far more line of sight. EM waves are used to carry information: The higher the frequency, the more information can be carried. The information to be conveyed is effectively mixed with the higher transmission frequency. On reception, the transmission frequency is stripped away, thus recovering the required information. Governments in different countries allocate these frequencies to different industries and applications for their transmissions. In the UK, TV is transmitted in the region 470 MHz to 862 MHz. One reason why governments are keen to promote digital TV is that the transmission mechanism is very efficient and so when digital takes over from analogue there will be spare frequencies that can then be sold off for use in other growing areas such as mobile phone.

5.2   Digital TV transmission requirements

Digital TV was designed with certain criteria that had to be met. These are as follows:

(i)   Similar bandwidth as existing analogue transmissions (6/7/8 MHz bandwidths)

(ii)   No interference with analogue transmissions when broadcast simultaneously

(iii)   Transmissions with bit error ratios (BER) of 10^{-4} (prior to error correction)

(iv)   Ability to cope with reflections

(v)   Optimisation to allow transmission from existing transmitter sites

(vi)   Local and national coverage

(vii)   Operation with single frequency networks (SFN's)

(viii) Reconfigureable hierarchical system such that capacity can be traded off against coverage in the deployment phase.

Point (i) refers to the fact that analogue TV uses an 8 MHz wide bit of the EM frequency spectrum for each channel. It was decided that digital TV would use a similarly wide range of frequencies. This is known as the bandwidth of the transmission. *Figure 5.2* shows this:

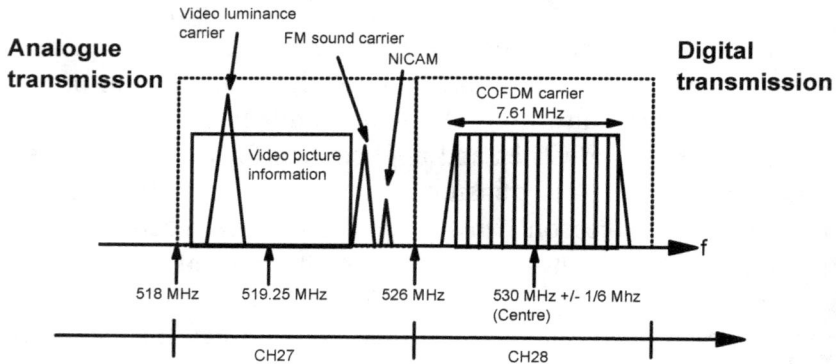

Figure 5.2

The diagram shows an example of an analogue channel such as BBC1 being transmitted in the 8MHz bandwidth of channel 27, at 519.25 MHz centre frequency. If the digital COFDM signal is transmitted on the next channel, *point ii)* requires that the COFDM signal does not cause any interference problems with the analogue one. Notice that the Nicam signal is at the far end of the analogue transmission. Any slight overlap of the digital signal could interfere with this. Hence an offset of +/- 1/6 MHz has been defined as part of the DVB-T digital TV specification. This is to allow the digital signal to be moved one way or another in case any interference is actually seen when the system is up and running.

Point (iii) states the number of errors that the channel can accommodate. A bit error ratio (BER) of 10^{-4} means that only 1 erroneous bit in every 10^4 can be tolerated. This can be understood by considering that the data transmitted is first very much compressed This (MPEG-2) compression, is of the order of a factor of 100. Also there is much format and control information that must not be corrupted. The error correction mechanisms are looked at in section 4.0 of this book, but it is their job to achieve very much lower error ratios from the incoming stream, thus producing the quasi error free (QEF) channel.

Points (iv), will be looked at in more detail once the COFDM principles have been covered in greater detail. However, clearly to be able to maintain a good BER reflections must be considered. It is reflections from other buildings and larger objects like mountains that give rise to the ghosting effects on analogue TV's at present.

Clearly as stated in **point (vi)** local and national coverage must be accommodated so that regional programs can be transmitted, as with the current analogue system. This has implications on the system design as regional transmitters of say 60 Km apart must not cause interference: A point talked about in more detail under section 5.4 on guard intervals. The requirement stated in **point (vii),** ie to operate with a single frequency network (SFN) is one of the most important in terms of defining the parameters of the COFDM network. The SFN requirement is an obvious one in terms of smaller and simpler frequency band allocations. The last point, **point (viii)** allows for the first phase roll out to be non hierarchical, ie simply to get a broadcast system up and running, but then later to be able to use the two level hierarchy scheme to transmit a standard stream of relatively low bit rate programming to achieve a high coverage, and at the same time to transmit higher bit rate, ie HDTV, to areas where reception is better.

Although mobile use was never a requirement set up by the DVB-T it is becoming more of an important area for not just video and audio broadcast, but also for data broadcast to mobile units.

5.3   OFDM frame structure

The basic structure of what is known as an OFDM signal is a variable number of frequency carriers, either 1705 (known as *2k mode*) or 6817 (known as *8k mode*). These carriers are spaced in such a way as to allow them to fit into the 7.61 MHz band width designated, as shown in *figure 5.3*.

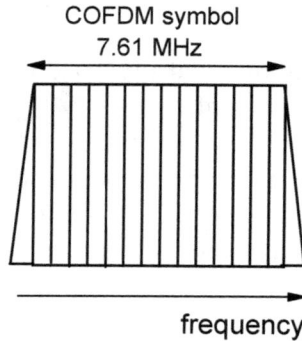

COFDM symbol
7.61 MHz

frequency

Figure 5.3

These carriers can be either:

(i)   **Data** - with a variable number of bits per carrier (2,4,6)
(ii)   **TPS** (Transmission Parameter Signalling) - transmission information
(iii)   **Pilot** - For receiver synchronisation, there are two types both transmitted at boosted power levels:
         (a)   **Continual** - 177 in 8k mode, 45 in 2k mode. These always at the same frequency.
         (b)   **Scattered** - 524 in 8k mode, 131 in 2k mode. Specified insertion pattern within the symbol.

This is shown graphically in *figure 5.4*.

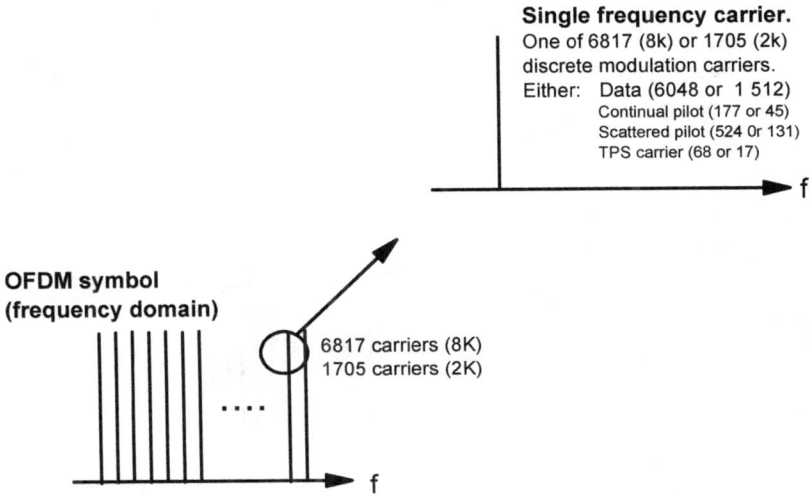

Single frequency carrier.
One of 6817 (8k) or 1705 (2k)
discrete modulation carriers.
Either: Data (6048 or 1 512)
Continual pilot (177 or 45)
Scattered pilot (524 Or 131)
TPS carrier (68 or 17)

**OFDM symbol
(frequency domain)**

6817 carriers (8K)
1705 carriers (2K)

Figure 5.4

Figure 5.4 shows the discrete frequencies that make up an OFDM signal, and so is shown by plotting amplitude against frequency. However, to get these frequencies to the receiver they must be transmitted over time. So lets now look at the time domain. The carriers are transmitted altogether for a particular length of time, this is the useful data time (T_u). A certain proportion of this useful data is repeated, and transmitted before the useful data. This is known as the *guard interval* and has two important functions (correlation and echoes). These will be looked at in detail later. This is now known as the *OFDM symbol*, and is shown in *figure 5.5*:

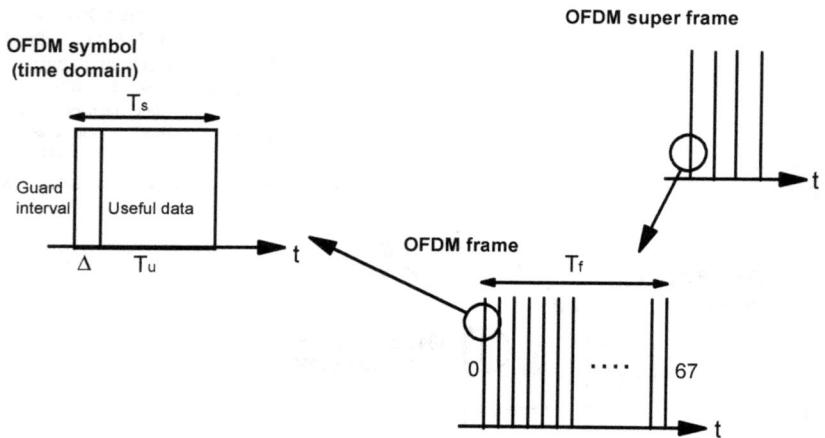

Figure 5.5

As can be seen from *figure 5.5*, the OFDM frame is made up of 68 OFDM symbols, with 4 OFDM frames making up an OFDM super frame.

5.4   The guard interval

The guard interval is a replication of the end of the symbol and is added to the beginning of the symbol. There are two main reasons for insertion of a guard interval:

(i)   To counter echoes and reflections.
(ii)   To allow the receiver to identify the start of a symbol.

These will be discussed in more detail in the next two subsections.

5.4.1   Echoes and reflections

The guard interval is a way of limiting the interference between the true signal and the reflected one. Specifically it counters what is known as *inter symbol interference* (ISI). This could be from a nearby building or mountain.

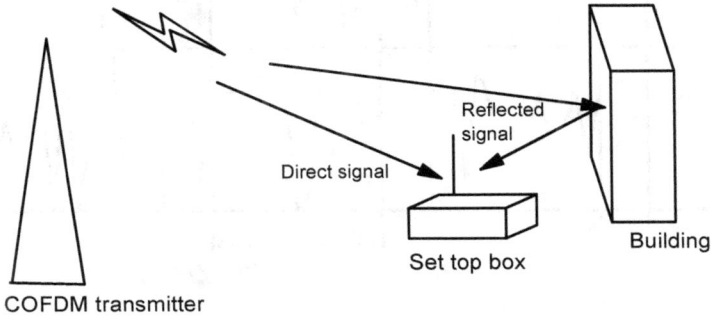

Figure 5.6

It also guards against situations where the same signals are transmitted at the same frequencies from different transmitters. This transmission architecture is what is known as a *single frequency network* (SFN). This opens up the possibility of also re-transmitting the same signal if the level is too weak at a particular location. A technique quite impossible with conventional analogue broadcast networks. It is the requirement that the system operate as a SFN from existing transmission sites that has driven the definition of the main parameters of the OFDM system.

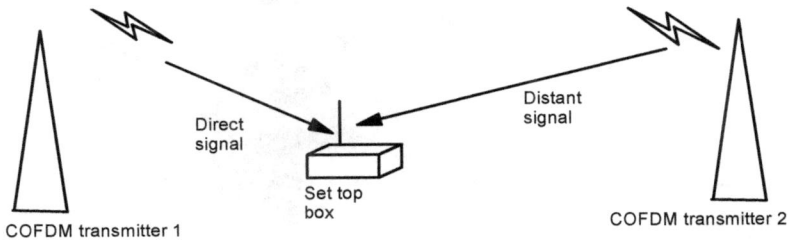

Figure 5.7

Figure 5.8 shows this in the time domain:

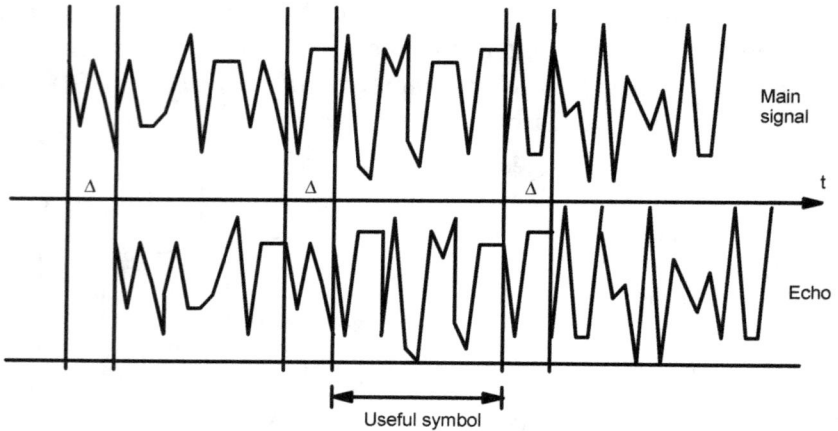

Figure 5.8

So how large should the guard interval be? Clearly this depends on issues like: How far apart are the reflective objects? Are they mountains or buildings? If a SFN, then how far apart are the transmitters? Lets have a look at the distances involved with a particular guard interval. In the UK a 2K single frequency network has been adopted. It will therefore use 1705 frequency carriers. The times involved and the useful bit rates, with the various transmission options, are shown in *table 5.1*:

Mode	8K				2K			
Guard Interval	*1/4*	*1/8*	*1/16*	*1/32*	*1/4*	*1/8*	*1/16*	*1/32*
OFDM Symbol duration	896 uS				224 uS			
Guard interval duration	224 uS	112 uS	56 uS	28 uS	56 uS	28 uS	14 uS	7 uS
Total duration	1120 uS	1008 uS	952 uS	924 uS	280 uS	252 uS	238 uS	231 uS
Max transmission rate (QPSK) MB/s	8.7	9.7	10.3	10.6	8.7	9.7	10.3	10.6
Max transmission rate (QAM 64) MB/s	26.1	29.0	30.7	31.7	26.1	29.0	30.7	31.7

Table 5.1

Looking at the UK the COFDM total symbol duration is 231 µs, with a guard interval of 7 µs.

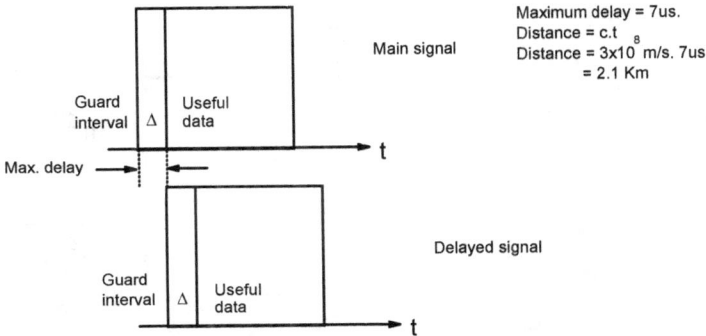

Figure 5.9

As can be seen in *figure 5.9*, before any inter symbol interference occurs the delayed signal has to be greater than the guard interval delayed in time. Since distance is velocity x time and the velocity is c, the speed of light (3 x 10^8 m/s), it follows that the delayed signal corresponds to a distance of 2.1 Km. This is fine in the UK where most of the reflections are off buildings and where local programs are being broadcast. However in mainland Europe and in particular in mountainous regions, and large national SFN systems, then this will not be enough. The DVB-T specification allows for a guard interval of ¼ in 8k mode. This corresponds to the maximum echo condition. ie 3×10^8 m/s x 224 µs = 67.2 km.

Having a large guard interval obviously has an impact on the efficiency of the bit rate and also on the number of carriers. Lets look first at the bit rate issue.

5.4.2   Guard interval impact on bit rate

Table 5.2 shows the useful bit rate (2k and 8k modes) as a function of the guard interval, the coding technique and the modulation type:

Modulation	Code rate	Guard interval			
		1/4	1/8	1/16	1/32
QPSK	1/2	4.98	5.53	5.85	6.03
	2/3	6.64	7.37	7.81	8.04
	3/4	7.46	8.29	8.78	9.05
	5/6	8.29	9.22	9.76	10.05
	7/8	8.71	9.68	10.25	10.56
16-QAM	1/2	9.95	11.06	11.71	12.06
	2/3	13.27	14.75	15.61	16.09
	3/4	14.93	16.59	17.56	18.10
	5/6	16.59	18.43	19.52	20.11
	7/8	17.42	19.35	20.49	21.11
64-QAM	1/2	14.93	16.59	17.56	18.10
	2/3	19.91	22.12	23.42	24.13
	3/4	22.39	24.88	26.35	27.14
	5/6	24.88	27.65	29.27	30.16
	7/8	26.13	29.03	30.74	31.67

Useful data rates in Mbits / s

Table 5.2

The bit rate can actually be worked out from the following equation:

$$\text{Useful data bit rate} = \frac{B \times C \times M \times N}{T} \text{ bits per second}$$

Where:

B is the efficiency of the RS block code (188/204 = 0.92).
C is the convolution code rate (1/2, 2/3, 3/4, 5/6 or 7/8).
M is the number of bits per carrier. (2 for QPSK, 4 for 16-QAM, or 6 for 64-QAM).
N is the number of data carriers used (1512 for 2k mode or 6048 for 8k mode).
T is the total duration of the symbol including the guard interval (eg 896µs + 28µs for 8k mode and 1/32 guard interval, or 224µs + 7µs for 2k mode and 1/32 guard interval).

5.4.3   Guard interval impact on number of carriers

The number of carriers is influenced by the guard interval required since there is a limit to the spacing of the carriers to maintain the orthogonality condition. (The orthogonality ensures that there is no inter carrier interference (ICI), ie interference of one carrier by its neighbour). This limit is in fact proportional to the inverse of the useful symbol length. Remember that the symbol is defined in the time domain, so this is a length of time. It may help to visualise this with the diagram of *figure 5.10*:

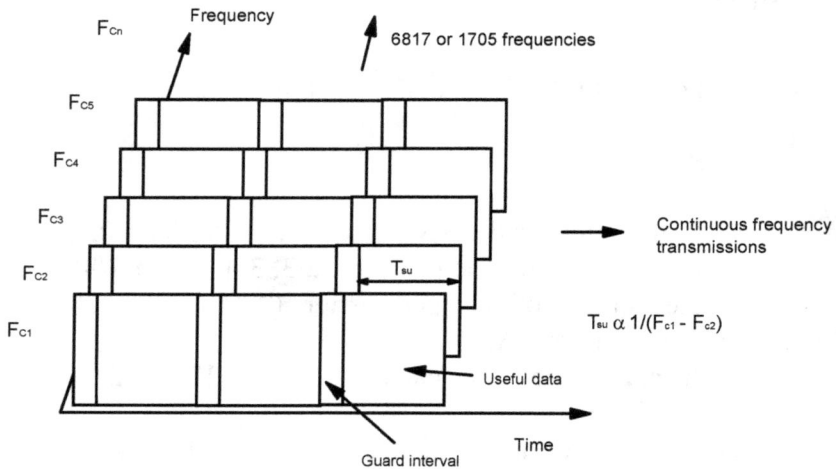

Frequency

F_{cn}

F_{c5}

F_{c4}

F_{c3}

F_{c2}

F_{c1}

6817 or 1705 frequencies

Continuous frequency transmissions

T_{su}

$T_{su} \propto 1/(F_{c1} - F_{c2})$

T_{su}

Useful data

Guard interval

Time

Figure 5.10

To maintain a reasonable bit rate it is accepted that the guard interval can't be more than ¼ of the total symbol duration. So to accommodate a guard interval of 200 µs (ie 60 Km path difference at the receiver from two different transmitters), with a total symbol time of 1 ms, the spacing of the individual carriers would become 1/ 800 µs, or an inter carrier spacing of 1.25 Khz. (ie the inverse of the useful symbol time). Since there is an 8 MHz bandwidth available, this means about 6000 carriers (8 MHz/1.25 Khz). The choice of the total number of carriers was a difficult issue. Some people were convinced that the selection of nationwide SFN's is key to the success of digital terrestrial television. Hence guard intervals of the order of 200 µs are needed, implying 6000 carriers per channel. OFDM is most efficiently implemented as an Inverse Discrete Fourier Transform (IDFT) at the encoder, and hence a Discrete Fourier Transform (DFT) at the receiver (see section 7.0 on OFDM theory). Since the technology used to calculate the Fourier transform acts on powers of two, this results in a DFT at the receiver of 2^{13} or 8192 (hence '8k') being the closest to 6000. However many people argued about the cost of such a complex receiver, making it non cost effective for the consumer. So a 50 μs guard interval solution was considered giving rise to a '2k' DFT size at the receiver. Finally the specification was defined for both systems. In fact today very little cost difference between the two systems at the receiver exists.

5.4.4   Symbol start detection

It was stated earlier that the guard interval performs two important tasks. The second of these is the symbol start detection at the receiver.

Because the guard interval is a repetition of the end of the useful data part of the COFDM symbol this should be detectable. The first thing a receiver must do therefore is to identify this and use this as an approximate method of finding out where the symbol starts. This is what is known as the correlation function and will be explained in more detail in part III of this book when talking about the receiver equipment.

5.5   The TPS carriers

The *transmission parameter signalling* (TPS) carriers are carriers that contain system information. There are 17 of these carriers per COFDM symbol in 2K mode or 68 in 8K mode. Only one bit of information is carried in each symbol, with all the information being carried over 68 consecutive symbols, or as previously defined, one COFDM frame. Each carrier per COFDM symbol carries the same bit (differentially encoded). *Figure 5.11* shows the 67 bits transmitted per frame, and their use:

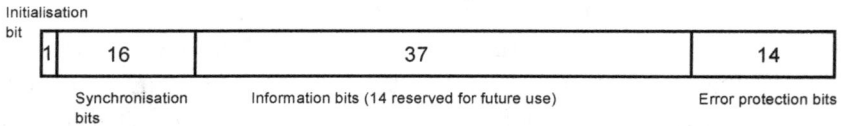

Figure 5.11

Of the 37 information bits only 31 are used at present, with the remaining 6 being reserved for future use.

The information conveyed to the receiver by the TPS carriers is:

(i)   Type of modulation scheme used including non-uniform constellation mapping information.
(ii)   Hierarchy information
(iii)   Guard interval
(iv)   Inner code rates
(v)   Transmission mode (ie 2K or 8K or 4k for DVB-H)
(vi)   Frame number within a super frame (ie 0 to 3)
(vii)   Cell identifier information for mobile use.

Note that these carriers are not modulated in the normal way as data carriers are. With data carriers there are various options available for modulation, ie QPSK and n-QAM. The TPS carriers are always differentially modulated, ie DBPSK. This is due to the robustness of this modulation scheme. This robustness is paid for by a reduction in bit rate. However this is not an issue for the TPS information.

The 31 information bits mentioned above are used as shown here:

Bits S_{17} to S_{22}:   length indicator
Bits S_{23}, S_{24}:   Frame number
Bits S_{25}, S_{26}:   Constellation

Bits S_{27} to S_{29}:	Hierarchy information
Bits S_{30} to S_{32}:	Code rate, HP steam
Bits S_{33} to S_{35}:	Code rate, LP steam
Bits S_{36}, S_{37}:	Guard interval
Bits S_{38}, S_{39}:	Transmission mode
Bits S_{40} to S_{47}:	Cell identifier

5.6   The pilot carriers

When COFDM signals are transmitted, there are a number of problems that can be introduced in the process. These, generally speaking, are:

(i)   Fading of certain carriers. This is a lack of amplitude. This can be caused by strong echoes causing destructive interference. This type of interference is termed *inter-symbol interference* (ISI). Note that constructive interference can also occur. The signal to noise ratio at the receiver can actually increase for certain carriers as a result of echoes. This power increase is actually an advantage, and with the use of good coding and interleaving techniques to correct for errors due to the fading, is the key that makes single frequency networks possible.

(ii)   *Co-channel interference* (CCI). This is interference caused by a transmitter in an adjacent channel, usually an analogue one.

(iii)   *White noise*, across all frequencies.

(iv)   *Adjacent channel interference* (ACI). This is caused, generally, in mobile reception conditions from an adjacent sub carrier. The carriers received from echo's or from another transmitter in a single frequency network should not drift to the point where they reach the next sub carrier frequency. In other words, in 8k mode, the Doppler effect should stay within 1 kHz.

This fixes a limit to the speed at which the broadcast can be received. This speed is also dependent on the carrier frequency.

(v)   *Phase noise*. This is when the phase of the modulated symbols varies with respect to the theoretical. This occurs as a result of inaccuracies in the receiver tuner local oscillator.

(vi)   *Doppler shifts*. This is a frequency shift error due to movements of the receiver, ie in mobile reception. It can also be introduced by movements of the transmitter mast.

The pilot carriers are there to estimate the amplitude and phase errors only. The results are interpolated at the receiver so that phase and frequency distortions for each carrier are available. This is known as *channel estimation*. These will then be used to correct for any errors that would otherwise be propagated. This is known as *channel equalisation*.

As has already been said, there are two types of pilot carriers; continual (45 or 177) and scattered (131 or 524). These pilots don't carry any encoded information, but do carry information in respect of their amplitudes and their positions. So let's have a look at the continual ones first:

5.6.1   Continual pilots

These carriers are always in the same place within the COFDM symbol, they are transmitted at increased power levels. These pilots are used to compute the common phase error (CPE). The CPE is an error that is introduced into the signal as a result of the local oscillator's phase noise. The local oscillator is in the receiver tuner. (See section 8.0). The CPE is a change in phase of all carriers within a symbol compared to the next symbol. We will look at the CPE later, and see how it is removed as part of the channel correction process.

5.6.2   Scattered pilots

These carriers, as the continual ones, are transmitted at boosted power levels. These pilots are inserted into the COFDM symbol in a predefined pattern, such that there is always a constant number of these per symbol (17 in 2k mode, 68 in 8k mode). These pilots are used in conjunction with the continual pilots to estimate the channel distortion.

5.6.3   Carrier visualisation

It is conceptually difficult to see what is actually being produced by the FFT . *Figure 5.12* below attempts to clarify the situation.

Data carriers

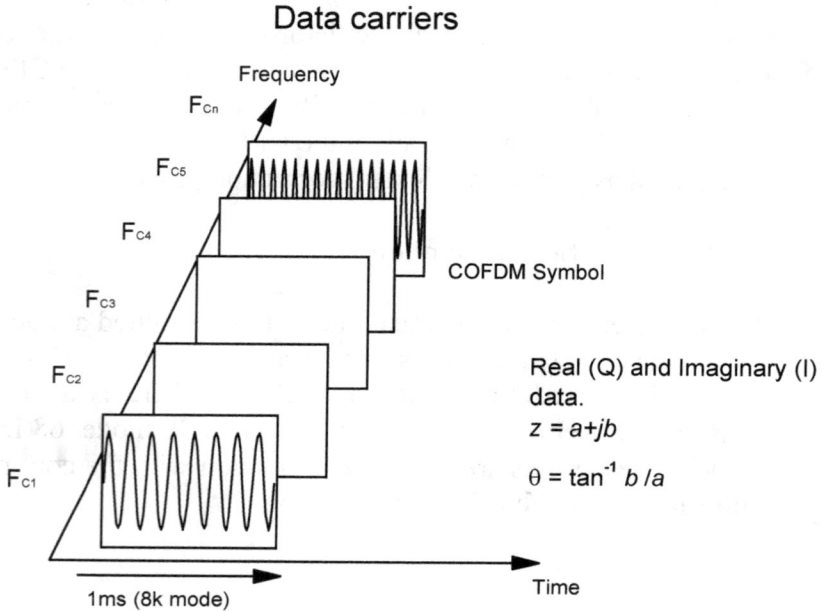

Figure 5.12

The diagram shows the data carrier frequencies propagating in time, as continuous waves (CW's). These contain the I and Q data by virtue of their phase and amplitude. TPS carriers are, as has already been said, differentially coded. This is shown in *figure 5.13*.

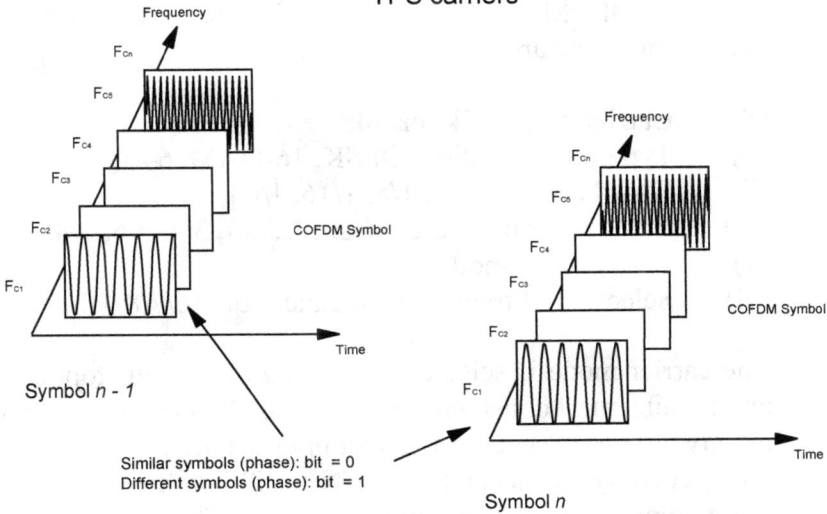

Figure 5.13

The TPS bits are transmitted using *differential bi-phase shift keying* (DPSK) modulation . That is to say that if the phase of a particular TPS carrier differs from one symbol to the next (by 180^0), then the bit is a '1', and if they are the same, then the bit is a '0'. There are 68 TPS carriers for 8K mode and 17 for 2k mode. The particular carriers used are specified in the DVB specification. Each TPS carrier within the same symbol carries the same bit information. The complete 68 TPS bits are defined over 68 symbols; an OFDM frame. The DPSK carriers in the very first symbol of a 68 symbol frame act as a reference and so do not carry useful data. Therefore 67 bits of data are carried by the TPS carriers in an OFDM frame.

5.7   COFDM Parameters discussion

It is clear that there are many parameters that can be adjusted for any COFDM based transmission system. To summarise, these parameters are:

(i)   Carrier mode: '2k' or '8k'
(ii)   Type of modulation: QPSK, 16-QAM, 64-QAM
(iii)   Guard interval: 1/4, 1/8, 1/16, 1/64
(iv)   Inner coder puncture rates: 1/2, 2/3, 3/4, 5/6, 7/8
(v)   Hierarchical modes
(vi)   Selection of transmission bandwidth (6/7/8 MHz)

The carrier mode is selected depending on various topological and architectural considerations, such as SFN and mountainous or city type landscapes. For example a larger guard interval will generally be needed for SFN's. '2k' rather than '8k' is more appropriate for mobile reception. This is a new and potentially massive market for data as well as audio/video broadcasts. The type of data and its required bandwidth will also have an impact on the parameters used. Ie high definition TV needing around 19 Mbits / second means that a high modulation type must be used and the guard interval and puncture rates can't be too relaxed.

The selection of modulation can also be made with the DVB-T COFDM system. The selection affects the data transmission capacity of a given channel, as well the robustness regarding noise and interference. The code rate of the convolution code (puncture rate) can then be used to fine tune the system performance.

When Hierarchical modes are used the data stream is split into high and low priority streams. The high priority stream is protected by a more powerful puncture rate, ie ½. This is also used to define where in the signal constellation the points are to be placed. The low priority stream is protected by a less powerful puncture rate, ie 5/6. This stream then modulates the low priority carriers in such a way that it defines a 'cloud' of constellation points around the points defined by the high priority stream. Hence in good C/N conditions each individual point of the constellation diagram can be identified and both the data streams can be detected. As the C/N reduces , and the 'cloud' points can no longer be detected, but simply in which quadrant of the constellation the 'cloud' is, then the high priority stream can still be recovered. This is shown in *figure 5.14.*

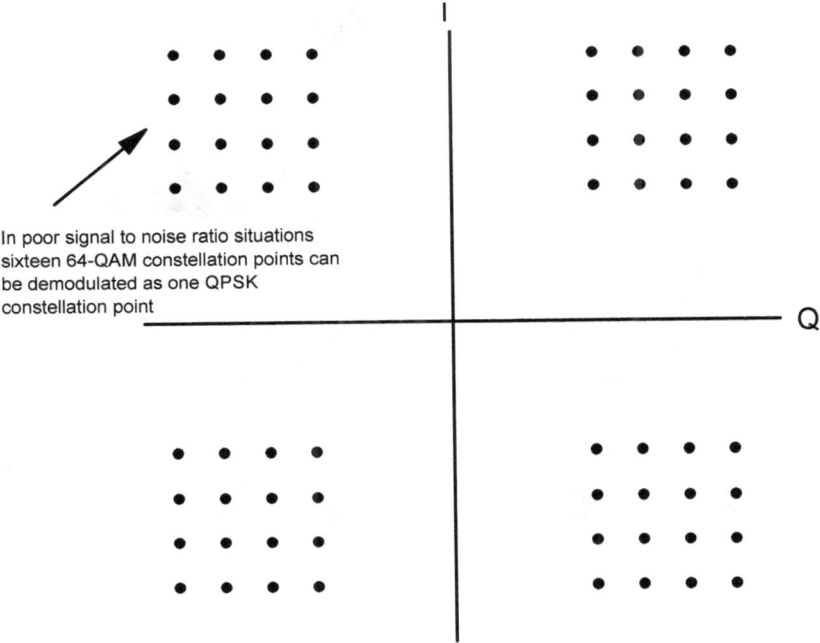

Figure 5.14

This could be useful, for example, in areas at the fringe of broadcast where deteriorating conditions can guarantee reception of a lower quality transmission with the higher broadcast being available when conditions improve.

DVB-T can also be transmitted in 8/7/6 MHz bandwidth environments with only the change of one oscillator frequency to allow the encoder to adapt to different channel spacings. For example a system clock of 13.5 MHz x 8192 / (858 x 19) should be used for a 6 MHz bandwidth system. This allows for a 6 MHz NTSC system to operate in a similar way to an 8 MHz PAL one. See reference [11] for more information.

6.0　　　　Overall transmission sequence

Figure 6.1 shows the basic functional blocks necessary to transmit a program stream using DVB-T COFDM techniques. Note the same blocks used for digital satellite broadcast (DVB-S).

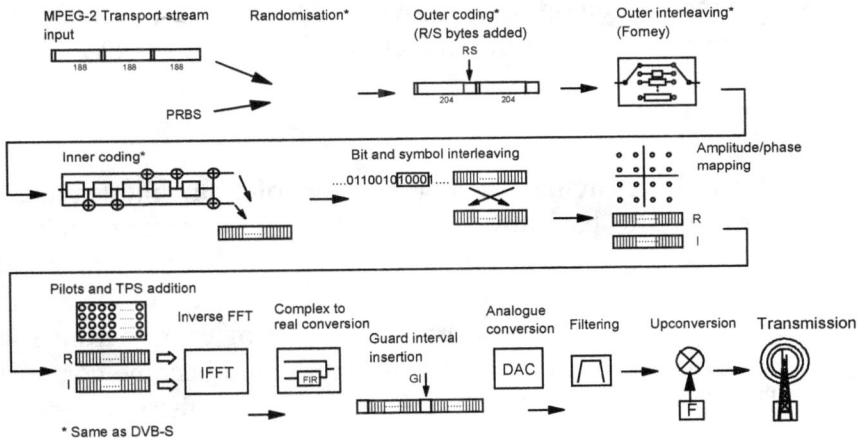

Figure 6.1

6.1　　　　Energy dispersal

As can be seen, the first step is to take the MPEG-2 transport stream. The stream could contain a large number of zeros or ones. This is unacceptable since it would introduce local DC levels within the signal, making both transmission and reception difficult. This possibility is removed by performing an energy randomising process. This can be better understood by looking at an impossible scenario: Consider if all the bits were zeros (or ones), then later on in the chain of events the guard interval would be added, but would be exactly the same as the data. hence the receiver would not be able to perform the correlation function and would therefore not be able to detect the start of a symbol. Details of how the energy randomising is achieved is given in section 4.1 of this book.

6.2　　　　Outer coding

This is applied to allow a receiver to work out which bytes have been corrupted during the transmission, and to correct for them. It can correct up to 8 corrupted bytes, and adds 16 additional bytes to the end of the 188 byte scrambled transport packet for this purpose. These 16 additional bytes are known as Reed-Solomon check bytes. This coding technique is described in more detail in section 14.4.3 of this book.

6.3　　　　Outer Interleaving

Outer interleaving or Forney convolution interleaving is performed to basically spread out the errors and so make the outer coding more effective. As stated above, the outer coding can correct up to 8 bytes in a 204 byte packet. Clearly if a burst error condition occurs, ie a burst of energy from some noise source, then more than 8 bytes could become corrupted. The Forney convolution effectively takes these errors and spreads them out over a number of packets, thus allowing the outer decoding to be more effective.

6.4　　　　Inner coding

This coding technique allows for the correction of errors not covered by the outer coding. It can be programmed to allow for more or less correction depending on how good or bad the transmission channel is. Again this is described in more detail in part I of this book. However it basically allows for the data to be repeated by a greater or lesser extent. Hence redundant data is transmitted. A puncture rate is defined, this, in a simplified way, can be considered to be the level of redundant data. The DVB-T standard allows for the following puncture rates: 2/3, 3/4, 5/6, and 7/8. A 2/3 puncture rate means that for every 2 bits of useful data 3 bits are transmitted , ie 1 redundant bit of data. So in a very clean channel 7/8 puncture rate could be used. In a very poor channel 1/2 would be used.

6.5   Bit-wise interleaving

The inner coder produces an output consisting of two streams maximum. The bit-wise interleaver produces 2, 4 or 6 streams (for QPSK, 16-QAM, and 64-QAM respectively).

6.6   Symbol interleaving

The purpose of the symbol interleaver is to map the 2, 4 or 6 bit words onto one of the OFDM carriers (1512 for 2 k mode or 6048 for 8k mode).

6.7   Amplitude / phase mappings

This maps the 2, 4 or 6 bit words onto the phasor diagrams as shown in *figures 6.2, 6.3* and *6.4*. Note the grey coding used:

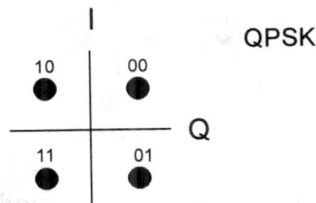

Figure 6.2

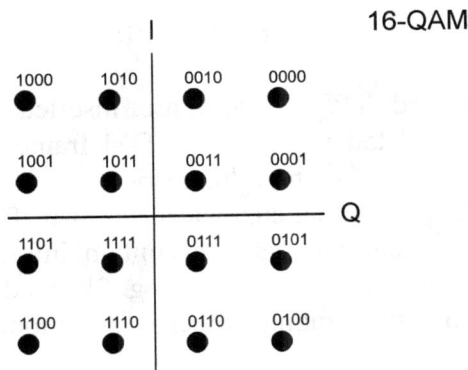

Figure 6.3

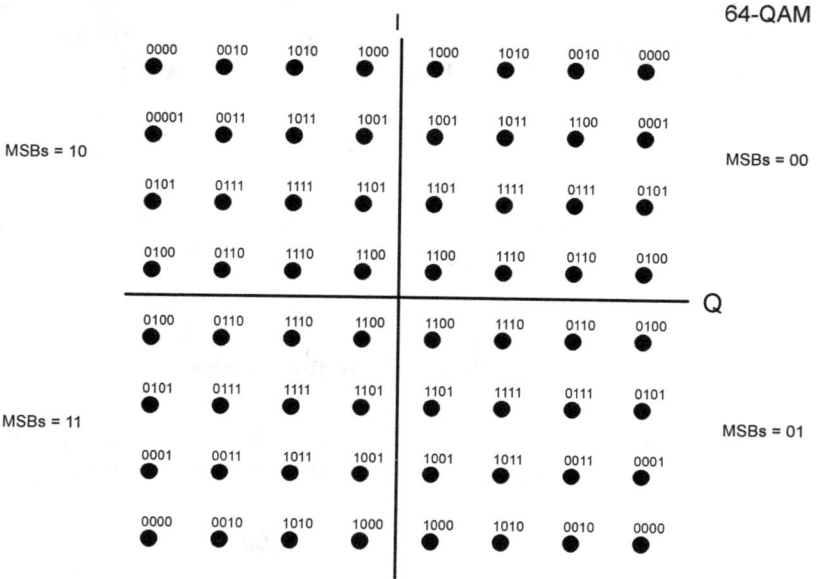

Figure 6.4

See section 20.0 of this book on modulation techniques for more information.

6.8   Pilots and TPS addition

The Pilot and TPS carriers are inserted next. These are described in detail under the OFDM frame structure section. But to recap briefly, the pilot carriers are used in the receiver to estimate the channel characteristics. The TPS (Transmission Parameter Signalling) carriers contain information regarding the detail of the data format, eg 2k mode, 64-QAM and constellation type, amongst other information.

6.9   Inverse FFT

This block effectively produces the 6817 (8k) or 1705 (2k) carriers. (Note that FFT's are done on powers of 2, hence 2048 or 8192). The outputs are added together to produce a single time varying signal. This is explained in section 7.1 on OFDM theory.

6.10   Time shift and combination

The time shift and combination section is responsible for generating the I and Q data. There are a number of different ways of achieving this. The DVB-T specification simply requires the I and Q data to modulate the carrier frequencies. However, this is typically achieved as shown in *figure 6.5*:

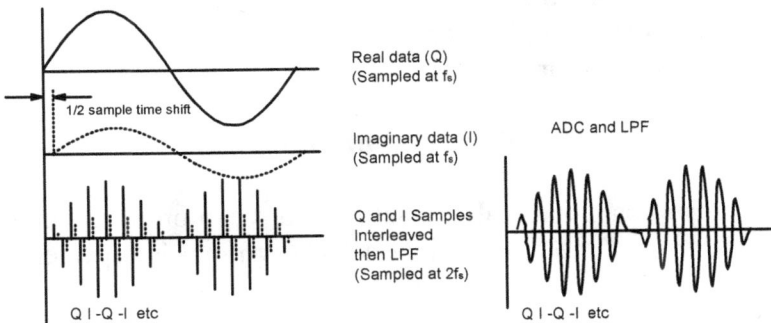

Figure 6.5

The output of the IFFT is 6817 or 1705 sine waves, each of different amplitudes. The I waves are added together, as are the Q waves, to produce two outputs; I and Q. Very irregular waves will normally result. However for simplicity *figure 6.5* shows these waveforms as time varying *sine* waves. The first two waveforms are therefore examples of what the Q (real) and I (imaginary) waveforms might look like in a simplified form. On output from the IFFT. One of these, the I data shown in *figure 6.5*, is delayed by a half sample time shift. Data is output from the IFFT at 9.14 MHz (f_s). This is the sampling frequency. After the I data is shifted, Q and I data are multiplied and added as shown below:

1) Q data multiplied by :   +1   0   -1   0

    etc.

2) I data multiplied by :   0   +1   0   -1

    etc.

3) (1) + (2):   Q   I   -Q   -I

    etc.

Hence when the addition is done the third waveform in *figure 6.5* shows the samples, which are now at an effective sample frequency of $2f_s$ or 18.29 MHz.

6.11   Guard interval insertion

Next the guard interval is inserted. To recap on this, a certain proportion of the data at the end of a symbol is repeated and added to the front of the same symbol. The two main reasons for this are:

1) To counter against echoes and reflections. (Primary reason)

2) To produce a correlation signal in the receiver, thus allowing it to identify the beginning of a symbol

6.12   DAC and filtering

The digital data is then converted to analogue and low pass filtered. The result of this is shown in the last of the waveforms of *figure 6.5*.

6.13   Upconversion and transmission

This uses standard techniques of mixing the time varying COFDM signal with a high frequency carrier. Again using the simplified sine wave example, this would simply look something like *figure 6.6* shows:

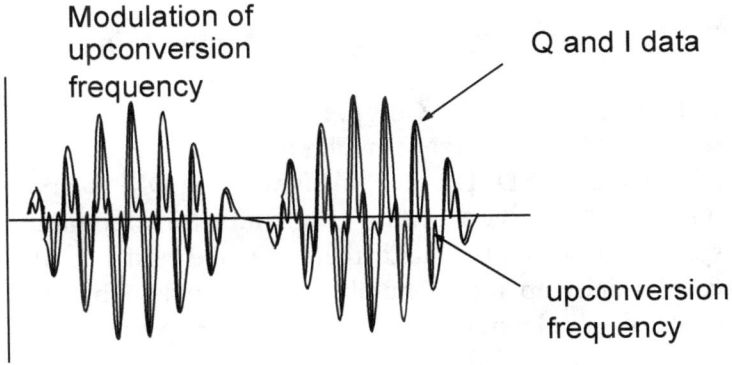

Figure 6.6

7.0   OFDM theory

The aim of this section is to explain how OFDM theory has evolved from a knowledge of Fourier methods. Section VII on standard theory explains the principles of complex numbers and Fourier transforms. It may be worth looking at this section first to recall these theoretical principles.

7.1   Frequency Division Multiplexing (FDM)

Before going into the digital world of OFDM it is worth looking at the analogue techniques of Frequency Division Multiplexing (FDM), since this is what OFDM is largely based on. If there exists a certain bandwidth for a communications system to be built around, then to make best use of this, the data is split up into a number of channels. This could be a number of telephone conversations for example, or a large amount of digital data that can be split for transmission, and recombined at the receiver. The data from these different channels is then modulated at different frequencies. Hence the data is transmitted simultaneously, making best use of the bandwidth available. *Figure 7.1* shows how a time varying signal can be split up into a number of channels:

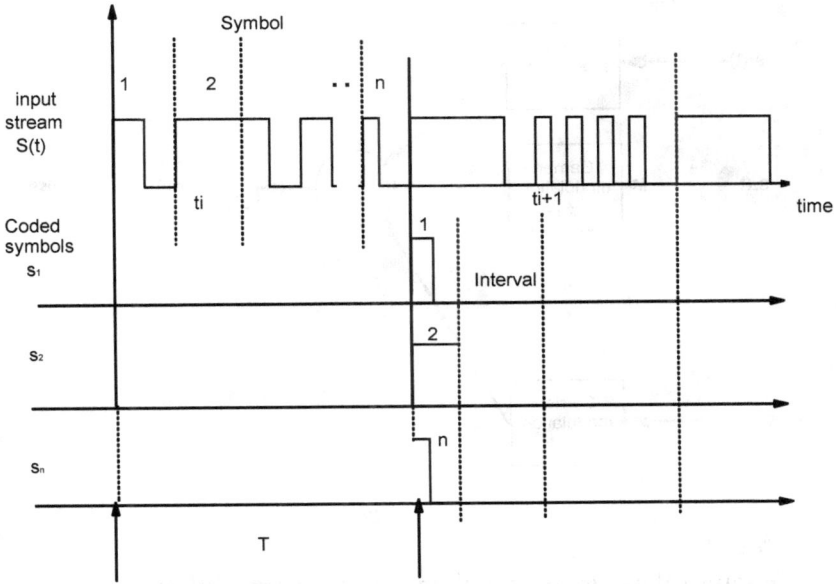

Figure 7.1

The main signal S(t), is split up into *n* symbols over a particular time period T. These symbols can be 2, 4, 6 etc bits wide, and are dependent on the modulation scheme used. (See section 21.0 on modulation techniques). *n* parallel data channels consisting of these symbols are then constructed. So we have produced *n* time varying signals from the original signal S(t).

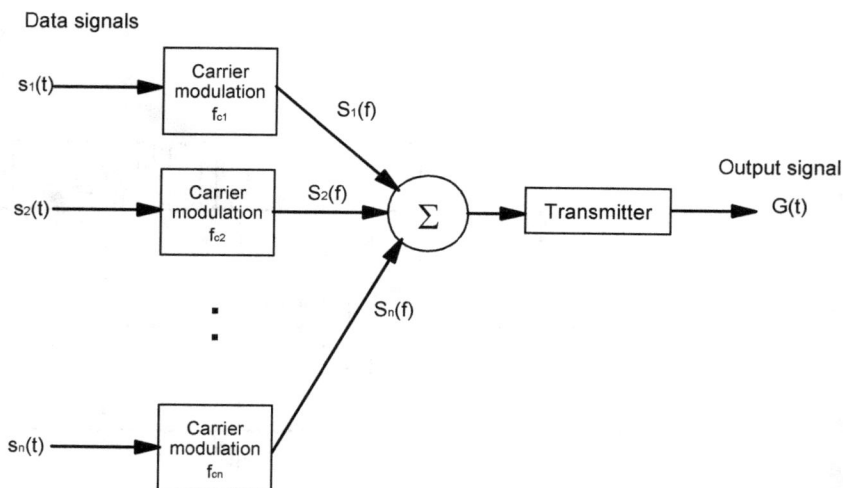

Data signals

s₁(t) → Carrier modulation f_{c1} → S₁(f)

s₂(t) → Carrier modulation f_{c2} → S₂(f)

Σ → Transmitter → Output signal G(t)

Sₙ(f)

sₙ(t) → Carrier modulation f_{cn}

Figure 7.2

Each of these signals is then separately modulated onto its own carrier frequency. These are then all added together and up converted to the transmission frequency. The final signal can be considered as a group signal, hence G(t); a signal containing all the frequencies of the sₙ(t) signals. This can be visualised as shown in *figure 7.3*, ie *n* symbols are transmitted for a particular time (T) in parallel, with different carrier frequencies.

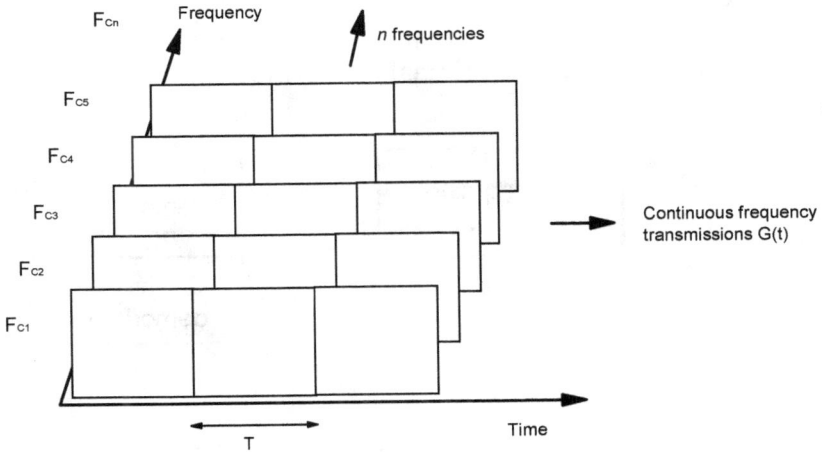

Figure 7.3

The time domain signal might look something like the waveform shown in *figure 7.4*.

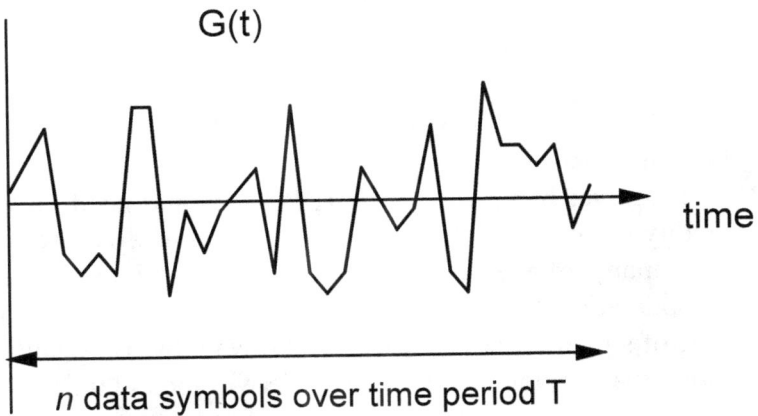

Figure 7.4

The complicated looking waveform is due to all the superimposed frequencies.

The receiver then accepts this waveform and processes it through an array of *n* band pass filters. These filters separate out the original carrier frequencies. The demodulation process then reconstitutes the original data symbols.

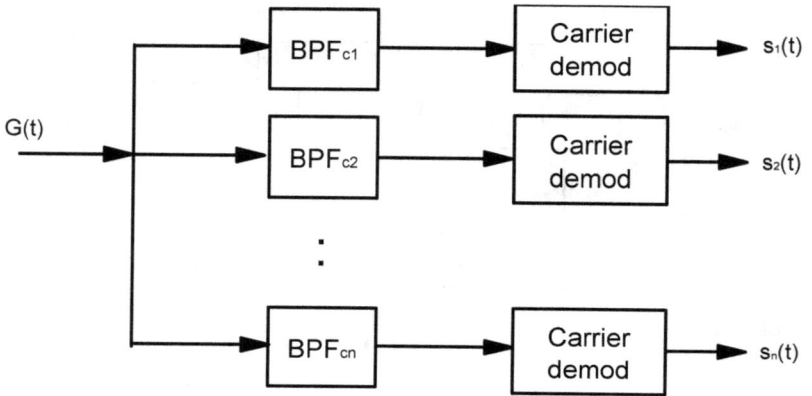

BPF = Band pass filter

Figure 7.5

This is the basic operation of a FDM system. However, to implement such a system when *n* is large is a costly business. The receiver band pass filters (BPF's) would need to be good quality to avoid interference effects, and a receiver would need very many of these. Also there would need to be *n* carrier demodulation blocks. The solution to this is to perform this type of signal processing in a digital way, by using calculation rather than analogue techniques. We will look at this next.

7.2   Digital FDM

Mr R. W. Chang worked out, in the late 1960's, that FDM was equivalent to an inverse discrete Fourier transform (IDFT) at the transmitter, and a discrete Fourier transform (DFT) at the receiver. This is shown in *figure 7.6*. Further more, he introduced the concept of orthogonality, which allows for the carrier frequencies to be close together, (thus making best use of the total bandwidth, in fact to almost the theoretical maximum), without experiencing either interchannel or intersymbol interference. The next sub section on orthogonality will explain these ideas in more detail.

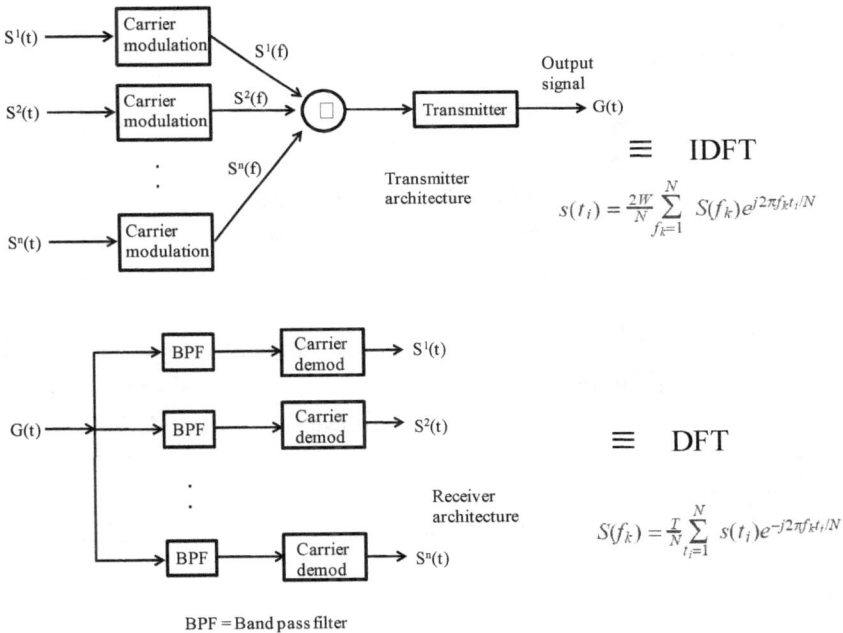

Transmitter architecture

$$\equiv \quad \text{IDFT}$$

$$s(t_i) = \tfrac{2W}{N} \sum_{f_k=1}^{N} S(f_k) e^{j2\pi f_k t_i / N}$$

Receiver architecture

$$\equiv \quad \text{DFT}$$

$$S(f_k) = \tfrac{T}{N} \sum_{t_i=1}^{N} s(t_i) e^{-j2\pi f_k t_i / N}$$

BPF = Band pass filter

Figure 7.6

The actual equation used for DVB-T COFDM transmissions is given in *figure 7.7* for interest:

Complex amplitude
Of active carriers

$$A(t) = \text{Re}\left\{ e^{j(w_0 t + \emptyset)} \sum_{m=0}^{+\infty} \sum_{l=0}^{67} \sum_{k=k_{min}}^{K_{max}} C_{m,l,k} \cdot \psi_{m,l,k}(t) \right\}$$

Upconversion
frequency

Frequency of active carriers

Where

$$\psi_{m,l,k}(t) = \begin{cases} e^{j2\pi \frac{K'}{T_u}(t - \Delta - l.T_s - 68.m.T_s)} & (l+68.m).T_s \le t \le (l+68.m+l).T_s \\ 0 & else \end{cases}$$

Figure 7.7

Where:

m	frame number
k	denotes the carrier number
l	denotes the OFDM symbol number
K	is the number of transmitted carriers
T_s	is the symbol duration
T_u	is the inverse of the carrier spacing
Δ	is the duration of the guard interval
ω_0	is the centre frequency of the RF signal (radians)
k'	is the carrier index relative to the centre frequency, $k' = k - (Kmax + Kmin)/2$

$C_{m,0,k}$

complex symbol for carrier k of the data symbol no. 1 in frame m

$C_{m,l,k}$

complex symbol for carrier k of the data symbol no. $l + 1$ in frame m

$C_{m,67,k}$

complex symbol for carrier k of the data symbol no. 68 in frame m

The summation from $l = 0$ to 67 defines the OFDM frame, $k = k_{min}$ to k_{max} defines the active number of carriers in the OFDM symbol. The real part of the whole equation is taken since clearly only the real part of the up converted symbol is transmitted.

7.3   Orthogonality

Orthogonality is the complete independence of one variable from another. The formal definition is:

Two vectors are said to be orthogonal if the cosine of the angle between them is zero.

For example two spatial dimensions are by definition orthogonal. Take for example a bullet fired from a gun. Imagine if there is a wind present, at right angles to the bullet's trajectory. It won't matter how strong this wind is, it can't cause the bullet to go faster or slower in the direction of its original trajectory. It will travel the same distance (again in the direction of its original trajectory) if the wind is there or not. The bullet's trajectory is therefore said to be orthogonal to the wind direction. But consider if the wind is at an angle. This is shown in *figure 7.8* in vector form.

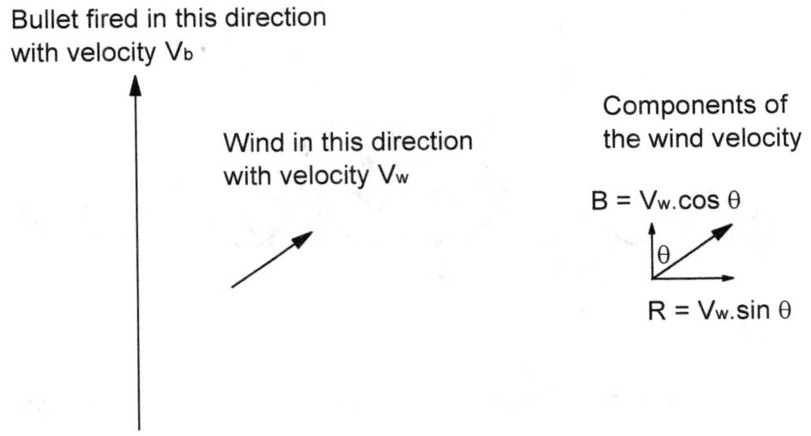

Bullet fired in this direction
with velocity Vb

Wind in this direction
with velocity Vw

Components of
the wind velocity

$B = V_w . \cos \theta$

$R = V_w . \sin \theta$

Figure 7.8

91

The wind is shown blowing the bullet to the right. There is now a component of the wind in the same direction as the trajectory. This is h.cos θ. So the velocity of the bullet now becomes $V_b + V_w.\cos\theta$. The bullet will now travel a little further than it would have if the wind were either not there or if it were at right angles to the trajectory. It can now be seen that if cos θ is zero there is no vector component in the trajectory direction.

In COFDM we are interested in interference effects. If signals are expressed as vectors, then to ensure that one signal does not interfere with another we must have these signals orthogonal to each other. This orthogonality is found to be true if the carrier spacings are such that they equal the reciprocal of the time period over which they are transmitted, as shown in *figure 7.9*. The shaping function must also be chosen such that the orthogonality condition is not damaged. The ideas of orthogonality are taken a little further in the standard theory section.

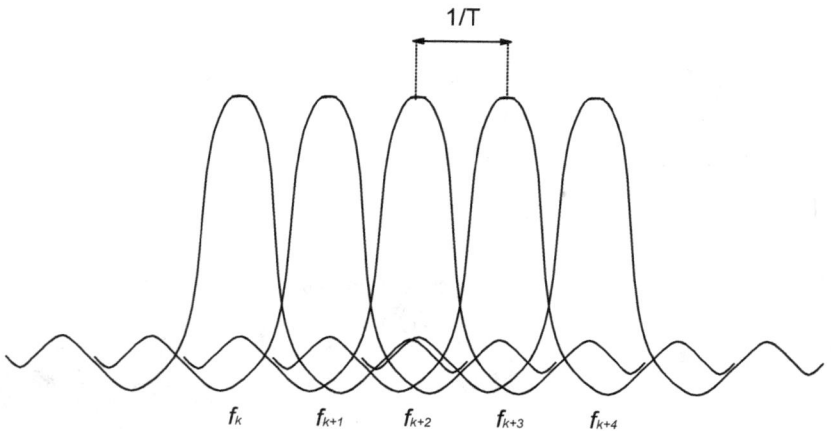

Carrier spacings = 1/T for orthogonality to be true.

Figure 7.9

7.4   Digital FDM functional blocks

Figure 7.10 shows the basic functions that are involved in a digital FDM system. As has already been explained, the various parallel time varying signals are input to the IFFT which performs the modulation process. The summed waveform is low pass filtered to ensure it fits neatly into its allowed frequency band. It is then up converted to its transmission frequency. See section 5.4.3 of this book for more information regarding selection of number of carriers.

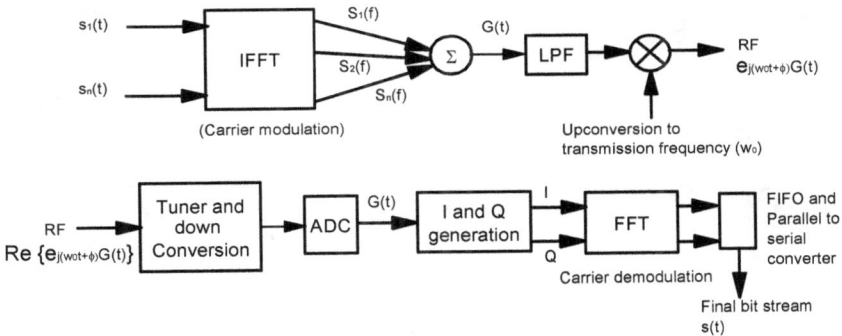

Figure 7.10

The signal is represented in its complex waveform, hence only the real part of the signal is transmitted. This refers to the real part of the up conversion frequency. After down conversion, the group signal $G(t)$ is reconstituted by sampling. The FFT effectively performs the demodulation. A FIFO is used for smoothing and a parallel to serial converter then allows the original bit stream to be output. The output stream is now a standard transport stream. These blocks will be looked at in more detail in the next part of this book on set top box architecture.

III   Set Top Box (STB) Architecture

8.0   Digital TV front end

The *front end* (FE) is the general term given to describe the electronics needed to take an input signal, be it from the LNB in the case of a satellite TV, the cable in the case of cable TV, or an aerial in the case of DTTV and to output the transport stream. In the case of DTTV, the front end, using the aforementioned definition, is made up of an analogue block and a purely digital block. The analogue block is referred to as the *analogue front end*. The digital block functions of a possible implementation are shown in *figure 8.1*:

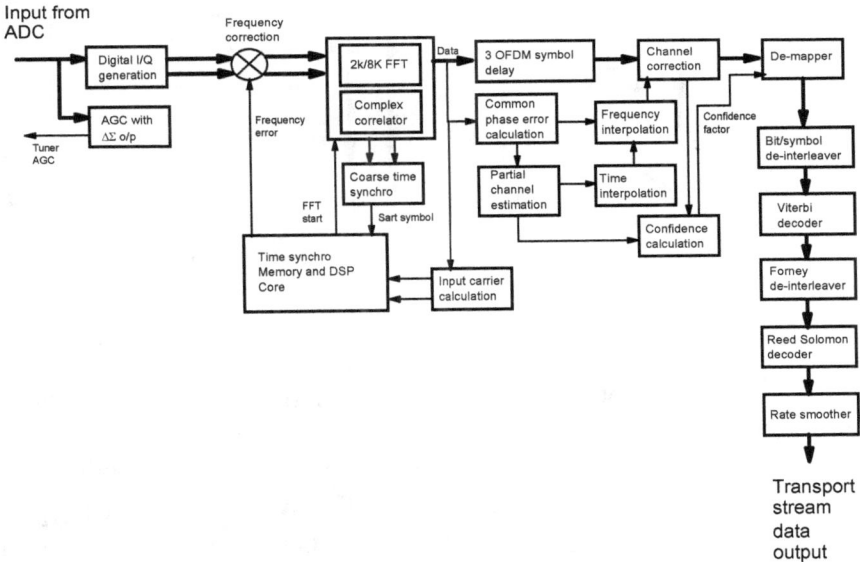

Figure 8.1

8.1   Analogue front end

The analogue front end consists of a tuner and an ADC (Analogue to Digital Converter) as shown in *figure 8.2*:

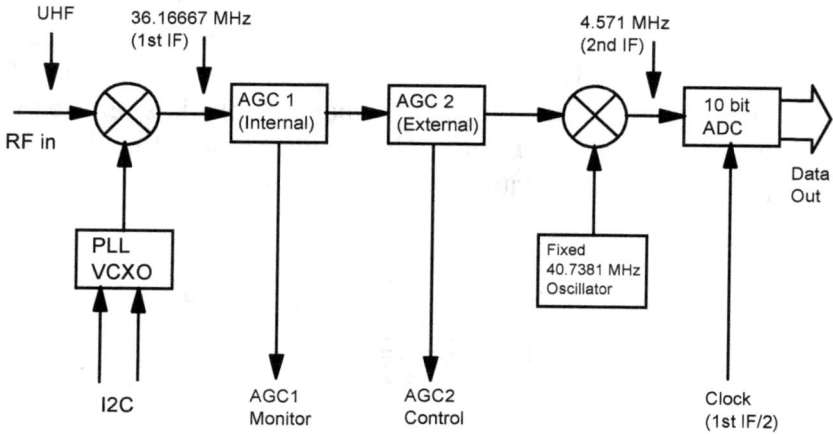

Figure 8.2

A tuner is needed in any TV system in order to convert the higher frequencies (used for modulation and transmission of the data) back to lower frequencies that can be used for recovery of the original data. The RF input is typically in the range 470 to 860 Mhz in a DTTV system. The PLL (phase lock loop) allows for a different frequency to be mixed into the incoming signal. It is this that is changed when you 'tune to a different channel'. Once this signal is mixed in via the heterodyne (shown in *figure 8.3*), a number of frequencies are produced. Hence a filter is needed simply to get rid of all the unwanted frequencies. The AGC (automatic gain control) blocks are used, firstly, to ensure that the signal amplitudes are the same for different input levels and secondly, to ensure that the A/D converter operates over its entire range. It is a balance between clipping of the signal and quantising noise.

In general, and also for digital terrestrial TV, the input frequency of say 500Mhz is simply the centre of a frequency band. In order to transmit the large amount of data a whole range of frequencies need to be used. The 500Mhz simply being in the middle of this band. If these frequencies are say between 496 MHz and 504 MHz then it is said that the bandwidth is 8Mhz. In fact 8Mhz is the bandwidth required for digital TV. (This is about the same as analogue TV except that this channel will practically contain five or six services encoded in the digital stream).

- F_c = 474 MHz (channel 21)
- PLL programmed to be 510.16667
- $F_I - F_c$ = 36.16667 MHz (Fixed 1st IF)

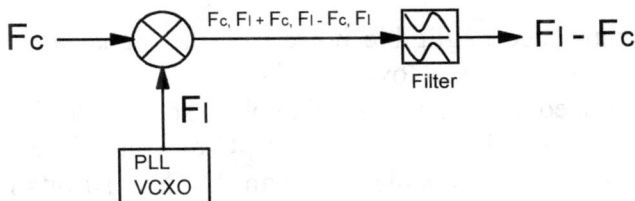

Figure 8.3

The heterodyne operation is shown in *figure 8.3*. As an example lets look at tuning into channel 21 at 474 MHz. To achieve this the PLL must be programmed to mix in a frequency of 510.16667 MHz. The output frequency therefore becomes 36.16667 MHz. This is known as the first IF (Intermediate Frequency). This frequency must always be 36.16667 MHz otherwise a single filter could not be used in the design, and stages further down the chain would be far more complicated and costly.

Going back to the tuner diagram, the fixed frequency oscillator (40.7381 MHz) is then used to simply down convert even further. The final outcome is to produce a stable single frequency signal at 4.572 MHz (32/7 MHz). This is then clocked out of the A/D converter at about 18 MHz. The data is then presented to the digital processing devices.

Many manufacturers are now supplying solutions whereby there is no need for the intermediate frequency. These systems are known as zero IF systems and can convert from between 950MHz to 2150MHz down to base band in one step.

8.1.2   The ADC (Analogue to Digital Converter)

The ADC clearly converts the analogue signal into digital. But how many bits are needed? And why is this clocked out at 18 MHz?

To introduce negligible noise into the signal an ADC of 7.5 bits is necessary. However, there could also be interference introduced from adjacent channels, so an additional 0.5 bits must be added. This gives an ADC of 8 bits. These must be 8 effective bits. To achieve this an ADC of 10 bits is generally used. (This is for a system that operates in 2K and 8K modes).

Why 18 MHz? First, the 18MHz isn't going to be exactly 18 MHz. This is the nominal value. It can be generated from a voltage controlled crystal oscillator (VCXO). This VCXO changes the frequency slightly to maintain synchronisation. The highest carrier frequency is 9Mhz. Hence we must sample at twice this frequency to adhere to the Nyquist sampling theory. (Note, although 6817 carriers are used in 8k mode, 688 dummy carriers are added to the left of the spectrum and 687 to the right. Frequency calculations are therefore performed using a true 8k definition of 8192). This highest carrier frequency is known as the complex sampling frequency (Fs). The 18.28 MHz nominal sampling frequency is therefore 2Fs. Incidentally the second IF is chosen such that the centre frequency is at 4.572 MHz since at this centre frequency the highest carrier frequency in 8k mode is at 9 MHz. This means that the sample frequency of 18.28 MHz is, as required, over sampling by a factor of 2.

8.2 Frequency and time synchronisation

The two main synchronisation functions that must be performed in a receiver are frequency and time synchronisation. The frequency synchronisation must be performed to compensate for errors introduced by the local oscillators used to down convert the RF signal to base band. The time synchronisation adjusts the nominal 18Mhz sample clock, ensures that the FFT window position is correct and performs the scattered pilots and TPS frame synchronisation.

8.2.1 Frequency synchronisation

Frequency synchronisation is performed prior to the FFT operation. One possible implementation is to split the process into two algorithms: Coarse and fine synchronisation. A final stage can then select and filter the correction signal with which the data can be corrected prior to being input to the FFT processing block.

8.2.1.1 Coarse frequency synchronisation

An algorithm can be used to act on the continual pilots to generate an estimate of the frequency offset of the receiver's local oscillators. This can be achieved to within roughly 1/3 sub carrier spacing. This can be done by initially achieving accuracy to within one carrier spacing, then improving this to within 1/3 of a spacing. Note that this algorithm can also be used to detect whether or not synchronisation has been lost.

8.2.1.2 Fine frequency synchronisation

Due to local oscillator phase noise, a complete OFDM symbol may be affected by a small shift, common to all carriers, but uncorrelated from one OFDM symbol to the next. This phase error is referred to as *common phase error* or CPE. A particular implementation is to estimate on each signal. This error can then be used to synchronise to within a fraction of a carrier spacing.

8.2.1.3   Selection and filtering

Again, there are other ways to perform this function. One application is therefore described here: This takes the coarse and fine outputs and decides, using a suitable algorithm, whether the receiver is to use the coarse or fine estimate for correction. The frequency error is then fed to an integrating filter and eventually to the sine / cosine tables where the values are modified. These tables are used to frequency correct the signal.

8.2.2   Time synchronisation

Time synchronisation is performed to control the ADC sampling clock, in order to achieve a good positioning and to avoid any movement or jittering of the FFT window signal. An FFT window signal can be produced to indicate to the FFT that it has the correct data and can proceed with the FFT calculation. To achieve this a correlation algorithm can be run. A particular implementation is described here: Since the guard interval is a copy of the last part of the COFDM symbol there is a correlation between two parts of a symbol. A signal can therefore be produced that identifies the position of a symbol. This signal can be used to drive a PLL that controls the frequency of the nominal 18MHz ADC clock.

8.3   Digital I and Q generation

The incoming signal now down converted and centred on 4.5 MHz (base band) can be filtered and converted to digital by sampling at twice the complex sampling frequency (18.285 MHz). This signal can be split into two signals at 9 MHz. One signal becomes the I signal and the other the Q. The Q signal can be filtered using an interpolation filter (16 tap symmetric digital filter), whilst the I signal can be delayed such that both signals can be delivered to the next stage at the same time.

8.4   Frequency correction

The I and Q data can now be frequency offset corrected. The frequency offset correction can be done in the time domain by multiplying the complex samples by $e^{j2on(\Delta f/N)}$, where n is the sample index and N is the FFT size. This can be achieved by using a complex multiplier and a table of synthesised sine and cosine values. Δf is the frequency offset. It is an offset introduced by inaccuracies in the local oscillator of the down conversion process in the tuner. The correction range can be of the order of +/-142 Khz, and +/- 572 Khz for 8K and 2K modes respectively.

8.5   OFDM demultiplexing

Figure 6.1 of section 6.0 of this book shows the overall transmission sequence. The decoding is basically the inverse of this with channel estimation and correction. The channel estimation and correction is performed after the FFT, with the demapping and subsequent digital decoding functions being performed afterwards. The FFT operation generally allows an 8K FFT to be calculated every 800 µs, and a 2K FFT every 200 µs.

8.6   Automatic gain control (AGC)

A calculation of the average input power is made, this is used as input to a first order delta- sigma modulator. The one bit output is used as the AGC. This is fed to the latter stages of the tuner circuitry to maintain a constant power level at the ADC input, thus keeping the digital output at a constant full range level.

8.7   Channel estimation and correction

The channel is the term used to describe the physical medium by which the transmitted signal is conveyed to the receiver. This channel will generally affect the signal in some way. How it affects the signal must be measured such that a correction procedure can be put in place. This is what is known as *channel estimation and correction*. The parameters that affect the signal are *common phase error* and frequency *response*. However, also part of channel estimation is to estimate the signal to noise ratio of each carrier. This is necessary in order to allow correct operation of the inner decoder (Viterbi). All these channel effects will now be looked at.

8.7.2   Common phase error (CPE)

Due to inaccuracies and noise in the local oscillator used in the tuner a phase error will generally be introduced into the signal. This may affect a complete OFDM symbol, but be completely uncorrelated between symbols. The estimation of this error must therefore be done on each symbol. The continual pilots are used for this purpose. What is actually done is that the continual pilots are used to calculate a phase shift for a particular symbol, this is repeated for a subsequent symbol, the difference is therefore a measure of the error. The phase error is generally denoted with the following symbol; δ_n. This is the common phase error between OFDM symbols n and n-1.

8.7.3   Channel frequency response

As the signal is transmitted across the channel, there may be some distortion in the carrier frequencies. Each carrier must therefore be corrected for this. However a picture must first be compiled to estimate by how much each carrier is likely to be out by. The scattered pilots are used for this. Calculations on these pilots and interpolation gives the required estimate.

The standard way of describing the properties of a communications channel is to measure how it affects frequencies of signals transmitted through it. A function known as the *transfer function* is defined that performs a modification on the original frequency. In the terrestrial channel we have:

$$Y_{n,k} = H_{n,k}.C_{n,k} + W_{n,k}$$

Where $Y_{n,k}$ is the demodulated sample of symbol n and carrier k. $H_{n,k}$ is the transfer function, or actual frequency response of the channel of symbol n of carrier k. $C_{n,k}$ is a point in QAM constellation space. $W_{n,k}$ is a 2D Gaussian noise source.

Now lets look at a method of estimating the channel: The estimate of the value of $C_{n,k}$ is noted as $C'_{n,k}$, and the estimated channel response is noted as $H'_{n,k}$. The frequency response is measured on the pilot carriers only, with time and frequency interpolation being used to estimate $H'_{n,k}$ for all n and k. The time interpolation is performed on four successive OFDM symbols (Hence a large memory is needed to allow for the storage of 3 OFDM symbols). Frequency interpolation is performed using a 23 tap FIR filter, although to limit hardware an 8 tap filter can be used with variable coefficient techniques. The $C'_{n,k}$ estimate is then performed by the following division:

$$C'_{n,k} = Y_{n,k} / H'_{n,k}$$

This is a critical calculation and can be achieved in two steps with the use of two multipliers.

These $C'_{n,k}$ samples now constitute the main signal that progresses through the subsequent processing blocks.

8.7.4   Signal to noise considerations

Each carrier may be affected by noise, or co-channel interference, adjacent channel interference or on board digital signals interfering with the analogue signal. These disturbances will generally have differing effects depending on the carrier frequency.

To add to this, frequency selective fading will cause some carriers to have a lower received power level than others. If you recall, selective fading occurs due to the destructive interference caused by echoes. The scattered pilots are used to calculate the signal to noise ratio (SNR). These are transmitted at boosted power levels for this purpose. In order to estimate the SNR for each carrier interpolation is used, hence memory is needed at the receiver to store these results.

8.8   Demapping

The demapping is the first of the decoding functions to be performed after the channel estimation and correction. The sequence of events is as shown in *figure 8.4*:

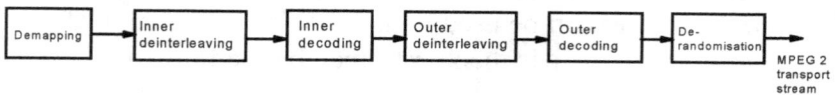

Figure 8.4

The demapping process converts the value of each noisy sample output from the FFT and channel corrector into a number of soft decision metrics, weighted by the channel state information. This is a confidence factor of the related sub carrier. Different metrics generation algorithms are used depending on the constellation used. This block produces 2, 4 or 6 metrics for a single modulation symbol.

8.9　　Inner de-interleaving

Inner interleaving was performed at the transmitter side to prevent a large notch in frequency selective fading causing long bursts of errors to occur. The bit de-interleaving receives the 2, 4 or 6 metrics from the de-mapper and feeds these into separate de-interleaver blocks. Hence there are 6 similar bitwise de-interleaver blocks.

8.10　　De-puncturing

Puncturing was a process performed at the receiver side by the inner coding block, or convolution encoder. It is a method of transmitting redundant data to varying extents. More redundancy is allowed in poor transmission areas, at the cost of a lower overall bit rate.

8.11　　Inner decoding

At the transmitter side a convolution code was used. This is a classical 64 state Viterbi decoder with soft decision inputs. A Viterbi decoder is therefore implemented at the receiver.

8.12　　Outer de-interleaving

At the transmission side the data was convolutionally interleaved on a byte basis to ensure that any errors are spread out. Therefore convolutional bytewise de-interleaving is required at the receiver at the output of the inner decoder. Data can then be fed to the outer decoder in the correct sequence.

8.13　　Outer decoding

The outer decoding is Reed-Solomon (RS204, 188) decoding. The coding produces 16 additional bytes to be added to the information data for error correction purposes. These allow for the correction of up to 8 errors. If the decoder fails in performing the decoding correctly a flag is set to signal this problem.

8.14   De-randomising

Another term for this is energy dispersal descrambling. The transmitter performed this pseudo randomising in order to disperse the energy, ie to prevent long sequences of ones or zeros, introducing a DC level to the signal. This de-randomising is simply the opposite of the generation process.

8.15   Byte rate smoothing

The purpose of this is to produce a constant rate data output. This is achieved by using a simple FIFO. The transport stream data output must not be too bursty or the demux processor will have buffer management problems. The PCR field in an MPEG transport packet contains temporal information regarding the packet. The error in the delivery of this packet should not exceed 500 ns.

8.16   MPEG-2 transport stream interface

This is the interface to the demux processor. This interface can be either serial or parallel. The parallel interface consists of 8 bits of data, a byte clock, a packet clock, and an error signal.

9.0   Digital TV back end

The *back end* (BE) is the general term used to describe the processing blocks after the recovery of the transport stream. The functional block of which is in *figure 9.1*.

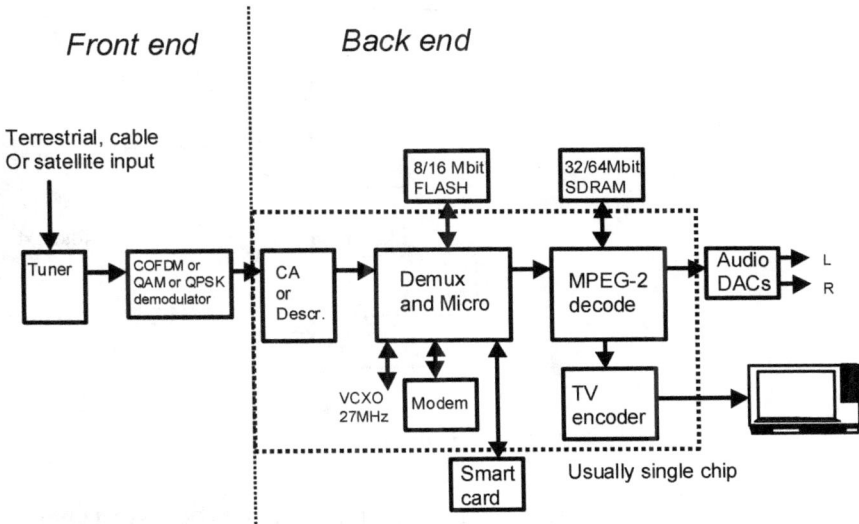

Figure 9.1

The transport stream is exactly the same for DVB-T, DVB-C and DVB-S, although the processing to produce it is quite different. Therefore the same back end can generally be used for all these applications.

9.1   Demultiplexing

Demultiplexing is the process of extracting all the useful information from the transport stream. For more information on the transport stream structure see part I of this book.

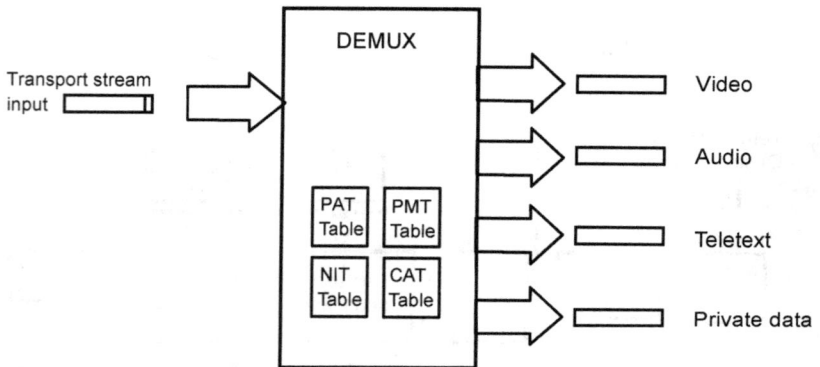

Figure 9.2

Figure 9.2 shows this function. The demux processor (Usually a dedicated 32 bit microprocessor) has a number of tasks to fulfil. There are:

1)   To construct the program association table (PAT), the program map table (PMT), the network information table (NIT), and the conditional access table (CAT). All these tables are described in part I of this book.

2)   To extract the compressed video data for a particular video channel.

3)   To extract the compressed audio data for a particular channel.

4)   To extract any other data required from the transport stream such as teletext data, conditional access

information, and any other private data such as a particular message.

In general the demux processor will not only perform the demux operation, but will also be responsible for running the entire set top box software, ie the basic operating system, the electronic program guide (EPG), various video control functions, interface to the front panel switches and status LED's etc. *Figure 9.3* shows the STMicroelectronics demux processor. This is based on the ST20 family of products.

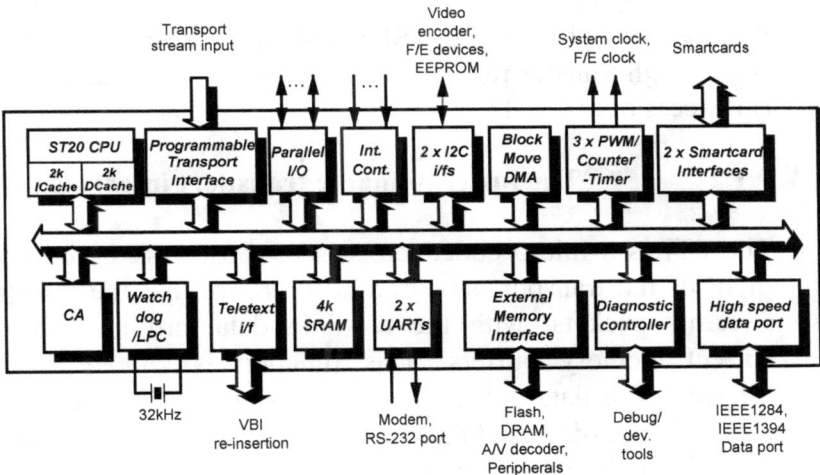

Figure 9.3

9.1.1   The ST20 CPU

The heart of the device is the ST20 CPU. This is a 32 bit variable length reduced instruction set computer (VL-RISC). A reduced number of instructions are used to simplify the internal architecture of the CPU and allow these instructions to operate faster. However in circumstances where more complicated instructions are needed, these are constructed. Around 70 % of instructions used are of the shorter (8 bit) and faster type, with only 30 % needing to be larger (16, 24 or 32 bits wide). This type of architecture also has the advantages of requiring less memory for program storage, and also offers speed advantages when accessing program code stored in external memory. The external non volatile memory will generally be organised as 16 bits wide, thus 2 instructions can be brought in on one external memory read cycle. Data and instruction caches are also included. These caches are fast internal memories that improve the performance of the whole system by copying data from external (slower) memory internally prior to the data actually being needed. However due to the performance of the *SDRAM* (Synchronous Dynamic Random Access Memory) devices STB's are generally designed with FLASH and SDRAM data interfaces of 16 bits wide. High transfer rates are achieved using a fast serial clock (in excess of 100MHz).

9.1.2 The programmable transport interface (PTI)

The PTI is a micro coded processor whose task is to accept input of the transport stream and to ensure that the transport stream's packets, with the required data, are put into the correct memory buffers. This function is known as *PID filtering*. The data will be put into various buffers depending on the nature of the data, ie video, audio, section data, teletext, conditional access or private data.

As part of a transport stream interface there will generally also be a piece of hardware to assist the software known as the section filter processor. This extracts the program specific information data contained in sections as defined by the transport stream specification.

9.1.3 Parallel input / output port

Parallel I/O pins are general purpose signals which can be used either to input data in the form of interrupt lines or a signal which can be polled by the CPU from time to time. These can also be used as outputs to light an LED or to send information to an external switching device for example.

9.1.4   Interrupt controller

This is a block of hardware that inputs interrupt signals from either an external device, such as the front panel microcontroller or switch, or from an internal device such as the UART. These signals tell the CPU that something needs attention. The interrupt controller decides what priority these inputs are and lets the CPU know that some piece of hardware needs attention. The CPU will then be scheduled to run the code associated with the particular interrupt needing attention.

9.1.5 I^2C buses

These buses are two wire buses. One wire carries a clock, the other the data. A protocol allows this architecture to transmit data in both directions, one device (The demux processor) must be the master. Each device that is connected to an I^2C bus has its own unique address. A number of external peripherals can therefore be connected to the bus at the same time. In a back end STB design two buses are used, and in general one bus is connected to the tuner, with the other bus connecting to the MPEG decoder, the digital TV encoder (DENC) and the front panel microcontroller. The tuner usually has a dedicated bus since noise must be kept to a minimum on the tuner interfaces to minimise phase noise problems. These buses are used to configure external devices and also for control. For example there may be a microcontroller on the front panel: The I^2C bus would, for example, be used to tell the main demux processor which infra red command it has just received, or the demux processor may want the front panel microcontroller to illuminate a particular LED.

9.1.6   Block move direct memory access (DMA) controller

The block move DMA controller allows large amounts of data to be transferred from one area to another with minimal CPU intervention. If, for example, some data for say the electronic program guide (EPG) needed to be transferred from the external memory to the MPEG decoder device for display, the CPU would give the DMA controller the address of where the data was, the address of where the data had to be moved to and the number of bytes the data consisted of. The CPU could then get on with some other tasks while the DMA controller performed the data transfer.

9.1.7 Pulse width modulation (PWM) counter timer

The PWM counter timer is used in controlling the voltage controlled crystal oscillator (VCXO) that clocks the whole back end system. Time stamp data is transmitted in the transport stream. The time stamp data is such that if the back end, including the MPEG decoder are clocked according to these time stamps there will be no overflow of buffers holding data to be decoded. Also the decoding will occur at the correct time. This is an important part of the demux operation and will therefore be explained in more detail:

9.1.7.1 PCR processing

The way it works is like this: The first received *programme clock reference* data is loaded onto a counter that is clocked down by the 27 MHz clock output of the VCXO. A new time stamp is then received. If the counter output has not gone down to zero then the value it has got down to is an indication of the error in speed. If a large value is found then the back end system is clearly not running fast enough. This error is then used, after some digital filtering, to modify the square wave that is being output by the PWM block. The mark space ratio will be increased. This square wave is externally filtered to produce a DC voltage level. With an increased mark space ratio this DC voltage will be larger causing the VCXO to increase in frequency by a small amount. A diagram of this is shown in *figure 9.4*

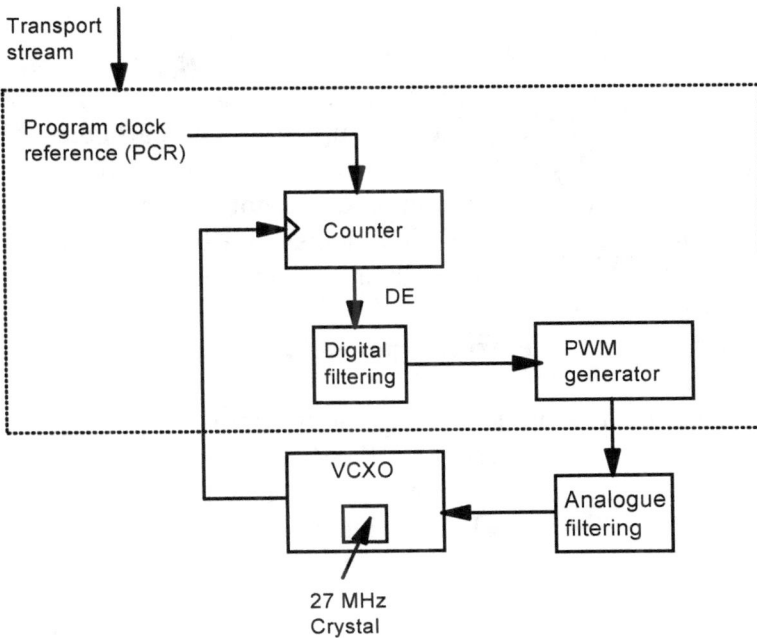

Transport stream

Program clock reference (PCR)

Counter

DE

Digital filtering

PWM generator

VCXO

27 MHz Crystal

Analogue filtering

Figure 9.4

9.1.8   Smart card interfaces

There are two smart card interfaces supported. One is generally used for the conditional access system. The other could be used for applications such as Mondex, allowing a viewer to pay for a particular program, event or service.

9.1.9   Conditional access (CA) block

An in built conditional access system allows the program provider to only allow people who have paid for a particular service to be able to decode the data associated with it. See the next section (9.2) of this book for more detail.

9.1.10   Low power controller

This block allows the demux processor to be put to sleep. An external crystal allows the block to keep track of time and can wake the demux processor up after a particular time. Alternatively it can be woken up if some activity occurs, such as the user pressing a button on the remote control unit. This is used depending on the various legislation being introduced to cut power consumption of electrical equipment.

9.1.11   Teletext interface

Teletext data is extracted from the transport stream and is sent to this block which serialises the data and passes this onto the digital encoder (DENC), which then inserts the data into the TV (PAL, SCAM, NTSC) signal for display in the normal way.

9.1.12   Static random access memory (SRAM)

There are a few Kbytes of internal memory on the device. Code and data accessed in this memory will be faster than that accessed in external memory. Time critical code and data is therefore placed here. This would normally be code associated with the PID filtering and the interrupt controller.

9.1.13   Universal asynchronous receiver transmitters (UART's)

UART's are used to realise an RS232 interface, and to interface with an external modem device.

9.1.14   External memory interface (EMI)

The EMI allows all external memory, and various peripherals needed in a back end system to be connected to the demux processor. Generally required to be interfaced is DRAM, FLASH memory and the MPEG video and audio decoder.

9.1.15   Diagnostic controller

The diagnostic controller is only used during system design when code is being written and tested. It allows code to be booted into the device. Once in and running it allows the code to be stepped through line by line, and individual variables accessed and changed. Various analysis functions are possible such as causing external triggers and timing code to see how long it takes to run.

9.1.16   High speed data port

The high speed data port allows data of any type to be output. This could be in the form of a transport stream or any other format decided on by the set top box manufacturer.

9.2   Conditional access

Conditional access gives the broadcaster the ability to restrict which viewers can and can't receive certain programming products. How this is achieved is very much dependent on the underlying encryption and scrambling technology. Broadly speaking, there are three main types of CA system available to the broadcaster, as shown in *figure 9.5*:

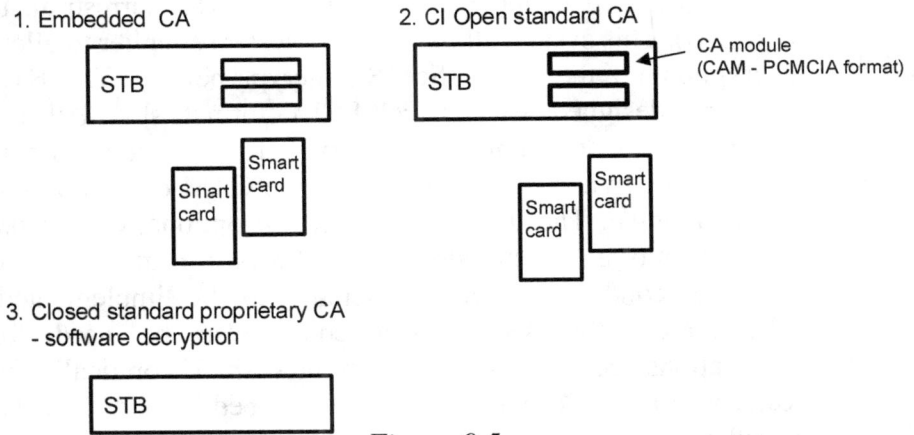

Figure 9.5

The first diagram, (1), shows the embedded system. For satellite, terrestrial and cable broadcasting, the main embedded CA providers at the writing of this book are:

- Nagravision, used in the US (Echostar), Canada (Bell - ExpressVu), Spain (Via Digital), Australia, New Zealand, Indonesia, Turkey, and Poland. In fact more than 40 broadcasters word wide.
- Viaccess, used in France, other areas of Europe, the Middle East, Asia and the US.
- MediaGuard for all Canal+ channels.
- NDS used in the UK (BSkyB), US, Spain, Israel, Australia and China etc.
- Irdeto used around the world.
- Conax in Scandinavia, Holland etc for DVB-T

The encryption algorithms are defined by the particular provider. A broadcaster typically pays a licence fee and ongoing royalty payments to the provider. A smart card must then be provided to the end customer to decrypt the encrypted channels. For this he or she typically pays a subscription fee or a fee for a single viewing in the case of pay per view (PPV) services.

The second diagram, (2), shows the *Common Interface* (CI) system (as used in the Europe and the US for terrestrial, ie NRSS-B). This system allows any CA system supplier to allow decryption on the STB. (See Simulcrypt below). The STB must be designed with a PCM/CIA format slot. All the electronics to descramble the encrypted transport stream is then supplied via a module known as the CAM or *conditional access module*. The transport stream can optionally be daisy chained into a second slot. This allows for at least two different conditional access systems to be implemented. Alternatively the second slot could also be used for applications such as Mondex electronic cash. Theoretically the same smart card can be used on an embedded CA or a CI based system.

The third option allows for many different ways of achieving encryption, from simple modification such as suppressing the PID values from the PAT or PMT tables, to more sophisticated systems. Such sophisticated systems, that use the standard DVB common algorithm, are available from the STB software design company Highgate Worldwide (www.highgate.tv) amongst others. These systems are ideal for the smaller cable companies who are more concerned with lower decoder box costs rather than extremely secure systems. Having said that it's worth noting that these *so called* secure systems are always prone to attack from hackers. At the writing of this book all the well known systems have been hacked with the exception of Seca 2 and Irdeto 2 (which, at least for now, are secure)

9.2.1   Operation of a typical CA system

To help explain how CA works lets first look at the method of encryption and decryption in STB's today. The data transmitted
to a STB is done so via digital transport stream packets. In a CA system this data is scrambled. However, also, as part of this digital data stream, certain information is not scrambled and is used to descramble the scrambled programming data. As is described in detail at the beginning of this book, the DVB MPEG-2 transport stream consists of many programs all multiplexed onto the same transport stream. This transport stream is then broken up into 188 byte packets for transmission: This is the *Transport Stream* (TS). However the individual programs are composed of many elements, such as video audio and text. The data associated with each program is also broken up into many packets prior to being multiplexed together with other programs. These packets are known as the *Program Elementary Streams* (PES). Generally it is easier to consider that encryption is the overall method of preventing someone gaining access to certain information, and that keys are encrypted and so need decrypting before they can be used. And the programming data is scrambled and must be descrambled before it can be decoded for viewing as part of

the overall CA system. The scrambling of the data can be done at either the TS or the PES level. (This is restricted to TS only as part of the ATSC specification for digital terrestrial TV). The basic way a STB decryption system works is as follows:

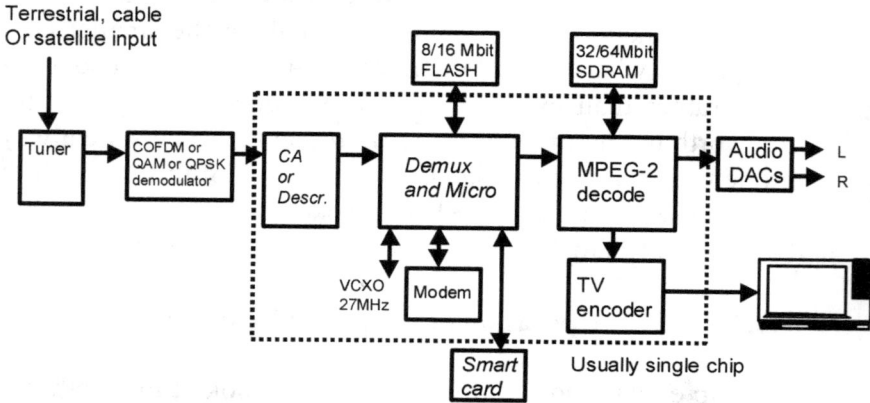

Figure 9.6

Figure 9.5 shows a typical set-top-box. The main areas that are involved with CA are the *CA or Descrambler* block, *the demux and micro* block, and the *smart card*. The *CA or descrambler* block shown might be a dedicated embedded CA module; for example in the case of BskyB (from NDS), or one of the standard descramblers, such as the US DES (Data Encryption Standard), or the European DVB one. Most broadcasters today use the DVB system from CA providers, such as Nagravision and Viaccess, or Conax. DirecTV are the only exception and use the DES standard (also from NDS). Let's now look at the mechanism involved.

9.2.2   The STB CA mechanism

The transport stream coming in from the air contains many packets of information. Each packet has associated with it (in its header) a *program identifier* known as the PID. All packets with PID value hex 1 are not encrypted, and are used, by the demux processor, to construct the *conditional access table*

(CAT). This table then identifies all the PID values of the transport packets containing *the entitlement management messages* (EMM's), More about these later. Also constructed by the MUX processor is *the program map table* (PMT). Again this is constructed from non encrypted packets and gives the PID values of all the transport streams associated with a particular program. Private data associated with the program can also be included in this table, ie the PID value of the packet containing the *entitlement control message* (ECM). The data contained in these two messages (EMM and ECM) are vital in descrambling the encrypted programming material. The mechanism of which will now be described:

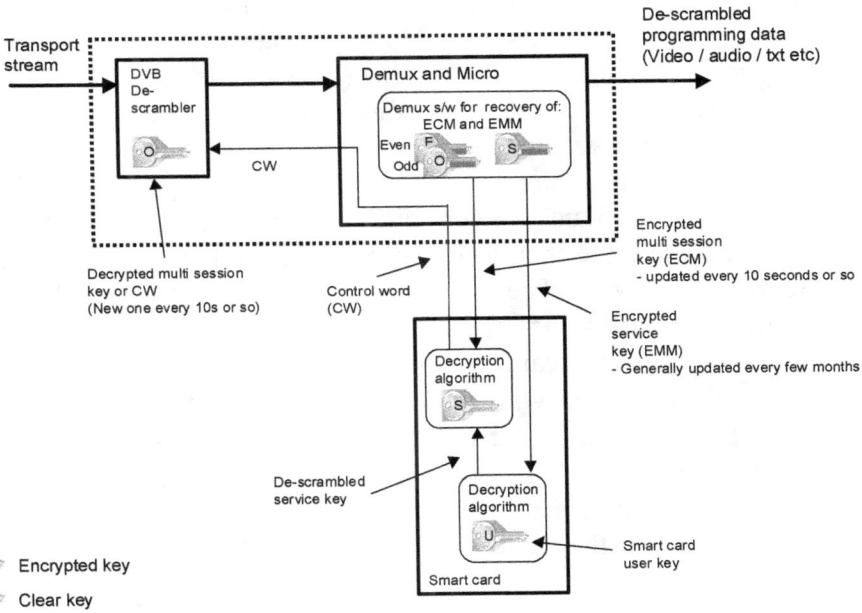

Figure 9.7

Figure 9.7 shows a DVB descrambling system. It should however be noted that the standards don't specify the smart card electronics or algorithms. The one described here is therefore a typical example. The EMM acquired by the DEMUX processor is related to the authorisation of services. It basically allows a particular set-top-box, or a particular geographic region to access services. It contains the *encrypted service key.* This key is typically changed every few months, so as not to give hackers too much time to decrypt it. The encrypted *multi session key,* carried by the ECM, is related to particular programming material. This key, once decrypted, actually becomes the control word that is input into the DVB descrambler allowing the transport stream to be descrambled so that the viewer can see a particular program or view the programming material for a particular session. As can be seen in the diagram, the service key (EMM) is sent to the smart card where it is decrypted with the help of the *user key* held inside the smart card. The descrambled service key is then used as the key to descramble the session key (ECM). The result of this is the *control word* (CW). It is this CW that is the key to the DVB transport stream descrambler. This is not quite the whole story: In fact there are two CW's. These are swapped every 10 seconds or so to make it very difficult for a hacker to crack the system. There are therefore two multi session keys giving rise to the two CW's. These are termed the *odd* and *even* keys. However, there is not enough time for the decoding of these keys in the smart card within the time allowed, and so two keys are decrypted with one being stored for future use. For example, the odd key derived CW may be being used, then, within the transport stream, a flag will indicate that the even key CW is to be used for the next section. The CA system will then swap over the control words as needed

As has already been said, the CA system can be (a) a totally proprietary system or (b) based on a common standard such as the DES or DVB descramblers. The DES system was developed initially by IBM in the late 1960's. The US National Institute of Standards and technology (NIST) since approved the DES descrambler for use over a certain time period. In fact the approval needs updating every five years. Last year the NIST announced a successor known as the *Advanced*

Encryption Standard (AES). This was selected to be the *Rijndael* system. However, the DES and the DVB systems are used today as standard systems for STB's. The DES control word is 56 bits long, but can effectively be extended to 112 bits by using two DES decryption blocks with two separate keys. In fact the ATSC specify a triple DES (TDES) mode for the CA system for terrestrial transmission. This 168 bit mode actually uses three single (56 bit) DES blocks with three keys. The first and last blocks perform encryption with the middle block performing decryption (on the transmit side). This is known as *high level MPEG security*.

9.2.3 Multicrypt and simulcrypt

There are different CA methods that can be used in STB's. The *multicrypt* method allows the CA system to be added as a separate module. The disadvantage of this approach is the additional cost. The DVB have recommended that manufacturers use the *common interface* (CI) for this purpose. This is a PCM/CIA format connector that allows a particular CA system (the complete system) to be 'plugged' into the STB, hence giving the capability for a STB to have any CA supplier system. Open Cable US systems use a point of deployment device (POD), which plugs into a particular STB for a similar reason.

Simulcrypt is a method that allows two CA systems to be used together. The broadcaster can transmit the ECM's and EMM's within the same multiplex to STB's with different CA systems. The broadcaster could also work with other multiplex operators
to increase his viewer base.

9.2.4 Advantages of CA to the broadcaster

As broadcasters look for more features to make their returns, the CA suppliers offer capabilities such as pay per view (PPV), where the smart card can act as a wallet with money being subtracted per movie or per minute. Messages can also be directed to specific STB's (maybe based on geographic region). Even certain movies can be forbidden to be watched, for example preventing kids from watching adult material.

Also more interactive features such as video on demand (VOD), games, etc are possible. The middleware selected is also of prime importance to the overall functionality of the system. The two main ones being used today are OpenTV (originally developed by Sun Microsystems and Thomson Multi Media) and Media Highway. NDS and Windows CE are also options. All these, of course, at a cost: Typically, a one time payment as well as royalties per box.

10.0   Silicon implementation

One of the most important driving forces behind the cost reduction and hence the take up of digital services is the silicon chip used to implement all the functionality explained in this book. A typical digital terrestrial set top box today uses around 3 or 4 major pieces of silicon. In the very near future these will all be incorporated into a single device. Some detail is given here to give the reader a better understanding of where this technology is today and where it is going.

10.1   The wafer

The main component in all chips is silicon. It is the 14th element in the periodic table, and is the second most abundant element on the planet. The most common form of silicon is as sand. In order for this to be useable there are a number of steps that must first be performed:
First the sand is purified and melted at a temperature of 1500 ^{0}C to produce a long bar of ultra pure grey coloured silicon about 8 inches in diameter. These bars are then sliced into very thin, about ½ mm thick, slices. These slices are known as *wafers*. The next step is the *oxidation* process.

10.2   Oxidation

The wafer is placed in a quartz oven at a temperature of about 1000 ^{0}C. The presence of water vapour allows oxygen to react chemically with the silicon, forming silicon dioxide (referred to simply as oxide). This oxide is chemically the same as glass, but extremely pure due to the contamination free environment used. The thickness of the oxide layer varies from 1×10^{-7} to 1×10^{-6} mm thick. The colour of the wafer depends on the thickness of the wafer. This oxide layer is a thin film that allows certain areas of the silicon to be protected from operations in subsequent processes.

10.3 Masking

A gelatinous light sensitive material, known as *photo resist* is now deposited onto the wafer. This behaves like unexposed photographic film. Between this coated wafer and a light source is placed the mask. The mask contains the circuit design and just like a negative is used to expose photographic paper, it is used to project the image of the circuit onto the wafer. The room where this is done is lit using yellow light since it is blue light to which the photo resist is sensitive. The areas that are exposed to the light are less resistant to a subsequent washing process than those that are unexposed. The washing process therefore removes the photo resist in the areas exposed to the light. Acids, usually in the form of a gas are used in the next process to remove the oxide that is no longer protected by the photo resist. The silicon underneath the oxide layer is however left undamaged. The remaining photo resist being no longer needed is then removed.

10.4 Doping

We now have a wafer that has silicon exposed in areas where we wish conduction of electricity to occur. The silicon therefore needs to be modified to allow this conduction. This is known as *doping*. This can be achieved in two ways. One method is *deposition and diffusion*: The wafer is inserted into an oven where Boron and Phosphorus are present in suitable forms. These elements impregnate the surface of the unprotected silicon. The second method is known as *ion implantation*: The wafer is placed in a particle accelerator where ions of phosphorous, Boron, Arsenic, Antimony and other elements are 'shot' at the wafer. This beam of ions penetrates the unprotected silicon modifying its physical characteristics making it conductive to electricity. Literally billions of ions are needed for each square centimetre of the wafer. To make these penetrate deep into the silicon the wafer is placed into an oven at $1000\ ^0C$ to $1200\ ^0C$ to allow the ions to spread out into the silicon.

The whole process may be repeated up to 24 times from the masking stage depending on the particular process used, using different doping elements and temperatures.

All the various circuit components must then be connected together. This is achieved using good conducting metals: Tungsten, titanium, aluminium alloys, silicon and copper. The various components can be transistors, memory cells, capacitors, resistors etc.

10.5   Testing

To ensure that the circuit performs as expected specialist test equipment is used with very small probes to connect to certain points on the circuit. Devices with any imperfections are then rejected.

10.6   Assembly and packaging

This step puts the chip into a package to both protect the circuit and to allow it to be soldered onto a final PCB. All the assembly operations are carried out in an environment where particles and humidity are controlled and manual work is reduced to a minimum to maximise the repeatability and reliability of the process. The first step in assembly is to cut the wafers onto a single dice. The chips are arranged on the wafer like a sheet of postage stamps and like stamps are separated along perforations; lines cut into the surface of the wafer. The cutting operation is performed with a diamond saw in a similar way as glass is cut. The chip is then attached to a copper support either by soldering with a lead-tin-silver alloy, or by using a glue loaded with particles of silver. This material ensures good heat conductivity. The support is known as a *frame* and has the connections (pins) that are used to connect the chip onto the final PCB. The next step is to connect the contacts on the chip to the leads on the frame. The numerous (more than 200) gold or aluminium leads are welded in a few seconds. Each wire is welded with micrometric precision to the chip and the frame completely automatically by a machine able to recognise the microscopic details of the chip using a TV camera and high speed image processing. The chip is now sealed in a block of thermosetting resin or in a small ceramic box. Protected against the elements, the packaged chip goes for forming and tinning of its leads.

10.7   Complete DVB-T implementation

The processing needed to realise all the digital COFDM functions (as shown in the block diagram of section 6.0 of this book), including all the error correction is of the order of 8 million transistors. With the 0.25 micron technology this can all be achieved within a single chip. Such a chip is of the order of 1 cm^2 in size with around 1W power dissipation. Note that 0.25 micron refers to the physical dimension required to realise a transistor ie 0.25 μm. Today these chips are implemented in 0.18 μm technology moving to 0.12 μm. The *back end* functionality is far smaller than the *front end* and is easily accommodated in a single chip using 0.5 micron technology, but also now available in 0.18 μm and with more functionality as detailed in section 18.0 of this book. A total DVB-T implementation (*front end* and *back end*) is actually possible today in a single chip.

10.8   System on a chip technology

When such complexity as the entire processing needed to perform the reception, decoding, demuxing, decompression, EPG, display etc. etc. of a set top box can be implemented on a single chip the term *system on a chip* is used. Semiconductors are no longer the low level building blocks that are used to build PCBs which are then used to build the final system. They are becoming themselves closer to the final system. In order for such systems to be produced the ingredients needed are:

(i) A good range of technologies capable of being integrated onto the same silicon.

(ii) Good process technology to give good yields at very low micron geometry's

(iii) Good broad range intellectual property (IP) and system software and know how.

Many companies are now producing SOC technological solutions.

11.0   STB software and MHP

11.1   Functional software architecture

At the start of DVB there was no real standardisation of software architecture within a STB. The functional architecture was shown in *figure 11.1*, this is still a good representation of the physical architecture in some STB's on the market today.

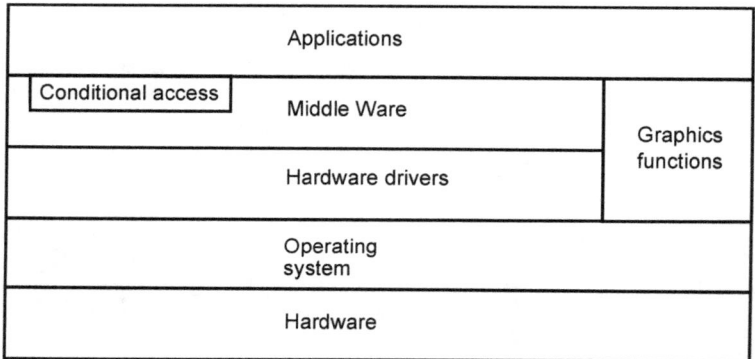

Figure 11.1

The hardware resides at the bottom, and includes all the standard components such as tuner, demux, MPEG decode, smart cards, conditional access, modem, scart switch, front panel etc. The operating system runs on the micro processor where the various drivers for the hardware reside. The operating system gives the drivers a good environment in which to operate, and will amongst other things allow them to share the CPU resource. It will also handle interrupts, scheduling and priorities. The hardware drivers allow simple commands to come in from the layer above; these are then translated by the driver into the low level commands and actions needed to perform the required task. The *middle ware* is generally service provider defined and gives a good environment for the applications layer software to run on. For example this layer may contain an engine on which an interpreted script at the applications layer will need in order to perform its function. The graphics functions are low level functions that can be called by the hardware drivers, or the middle ware for display purposes. For example, an electronic program guide (EPG) at the applications layer may need to put some text on the screen. A particular graphics function will perform all the low level line drawings necessary to allow the text to be seen at a particular size, font, position and colour. The conditional access system for digital terrestrial television is a software one (MediaGuard) and also resides in the middleware.

11.2   Software standards

The various software specifications are shown here as they relate to *figure 11.1*:

11.2.1   Operating systems

These are dependant to a certain extent on the type of processor used. Some processors may have a particular operating system ported to them, whereas some might not. The STMicroelectronics ST20 processor is the most widely used one for all STB applications to date. The following are all different types of operating system:

OS20, μCos, VxWorks, Windows, UNIX, Osek, linux, windriver, QNX, etc. However the main ones used in STB applications are OS20 and μCos.

11.2.2   Hardware drivers

These are very specific to the type of hardware used, and are generally, but not always specifically written by the silicon provider who knows the hardware best.

11.2.3   Middleware

The middle ware is the platform that the applications layer programs are written for, eg the EPG. These include: MediaHighway DLL 4.x, MediaHighway DLL 3.x, OpenTV1.x, OpenTV OpenTV EN2, WinCE. The standard adopted for digital terrestrial TV in the UK is MediaHighway 4.1.

11.2.4   Applications layer

The applications layer, is written by STB manufacturers or third party software houses to perform the various tasks required for the particular service provider, ITVDigital for the terrestrial service.

11.3   Actual Software architecture

Prior to MHP there was no real overall hardware independent architecture. Chip manufacturers, such as STMicroelectronics, were keen to have solutions come to market whereby the layers above the hardware drivers could be re-used with no major software code re-writes and not so be dependent on one particular device. Hence the situation became a little more complicated than the functional model explained above. *Figure 11.2* shows the STMicroelectronics architectural model.

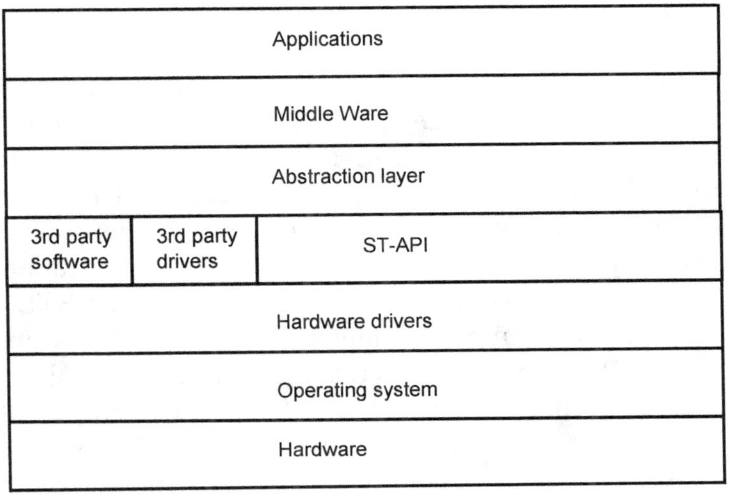

		Applications	
		Middle Ware	
		Abstraction layer	
3rd party software	3rd party drivers	ST-API	
		Hardware drivers	
		Operating system	
		Hardware	

Figure 11.2

The new layer shown here is the ST-API (Application Programming Interface) layer. This was written such that it offered a standard interface to the next layer up in the architecture. This allowed STB software writers to keep the same middleware layer (or optional abstraction layer) on all their boxes whenever the hardware, and so the hardware drivers changed. This allowed the STB manufacturer to take advantage of all the new hardware features of new devices with no additional software effort, making the final box quicker and cheaper to produce. The optional abstraction layer allowed the manufacturer to take account of any new hardware not in an integrated chip, or additional s/w such as TCP/IP for internet access.

11.4   MHEG-5

Firstly, what is MHEG. It is an acronym for Multimedia and Hypermedia Experts group, and is a standard developed within ISO (ISO/IEC 13522-5). MHEG-5 is simply the fifth part of the MHEG standards suit. It was developed to support the distribution of interactive multimedia applications in a server / client type architecture. By allowing an application to be distributed across both server and client (ie the STB) means that the STB can have a much smaller memory requirement. (An MHEG-5 engine, which is the part of MHEG that runs on the STB, is only of the order of a few hundred Kbytes of code). This engine can then be ported onto any STB with any core microprocessor and the same application will run in the same way. This is the main purpose for such an architecture. See section 12.9 of this book on interoperability.

The middleware that was used on the ITVDigital (UK digital terrestrial subscription system – now no longer in business) was MediaHighway 4.1 from Canal+ technologies. One reason for this choice was that they had a working MHEG engine. The MHEG engine allows interpreted script applications to be run. MHEG can be considered to be the HTML or Java/JavaScript equivalent for the interactive STB environment. MHEG-5 allows for such things as EPG's (for navigation), teletext functions and home shopping all to be run on the STB. The UK system today is a free view model with UK profile MHEG-5. An important aspect of this solution is the DSM-CC (Digital Storage Media Command and Control) protocol that is used to realise the data carousel. This is a user to user transmission protocol for objects. For example the UK teletext data is transmitted as such MHEG-5 objects. (See section 12.6.1 of this book for more information regarding DSM-CC)

12.0   MHP

The DVB have successfully developed many standards for digital broadcast, now used throughout the world. These primarily at the drivers level of the software stacks shown above. However, due to the convergence of technologies for reception at the home the DVB (under the encouragement of the European Broadcasting Union and the UNITEL project) initiated meetings, as early as 1996, to define the commercial requirements for future interactive receivers. This culminated in the DVB-TAM (DVB-Technical issues Association with MHP) ad-hoc group who started the MHP (Multimedia Home Platform) definition in 1997. Specifically the API. This then moved the DVB into the higher software stack layers and into the area of middleware and applications. The main driving force behind MHP was to offer interactivity to the end user. It was to specify the functionality of applications rather than to specify the complete middleware in absolute detail itself. The API's are well defined, but how these API's are put together in a real architecture are not. MHP version 1.0 was approved on the 23rd of February 2000. This specification defines the following two profiles:

P1)   Enhanced Broadcast - Enhanced middle ware features, ie digital broadcast of audio / video services including downloaded applications for local interactivity, but very limited interaction with just an optional telephone or cable modem return channel.

P2)   Interactive – Ability to run more interactive applications across the return path. ie a more sophisticated interactive channel is needed.

A third profile was then defined (and approved in June 2001):

P3)   Internet – Allowing HTML and other protocols to run, via a direct connection to the Internet.

This then completed the MHP middleware as a true interactive standard.

Profiles 1 and 2 are very similar and are specified in MHP version 1.0, version (iii) is specified in MHP version 1.1 and also contains versions (i) and (ii). So to summarise:

MHP ver 1.0: Enhanced broadcast (P1) and Interactive (P2)
MHP ver 1.1: Ver 1.0 + Internet (P3)

12.1　　　MHP interfaces

MHP has the capability to interface to:

i) Programme streams for input and output.
ii) Data stream sources, also for storage
iii) Presentation engine for data display to the user
iv) Viewer input devices for viewer interactivity

An example of an MHP system interfacing to the outside world is given in *figure 12.1*.

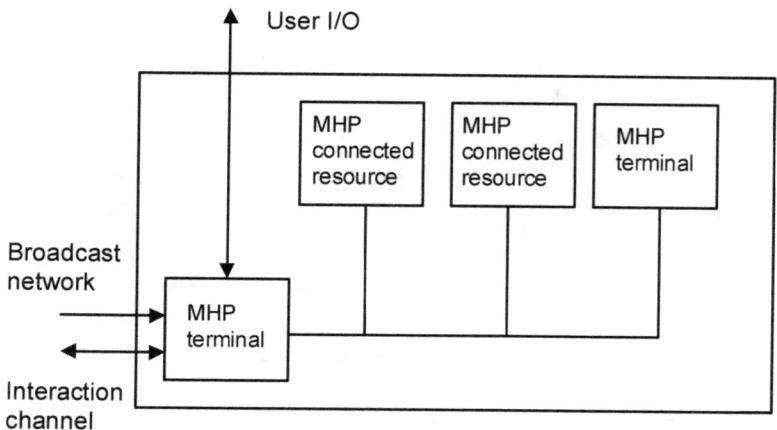

Figure 12.1

The *MHP terminal* is any piece of equipment that conforms to the MHP specification, and in particular has *a virtual machine* (VM) and an instance of the MHP API. *The MHP connected resource* is part of MHP but by itself does not conform totally to the MHP specification. So now we know the interfaces let's look in more detail at the architecture.

12.2   MHP architecture

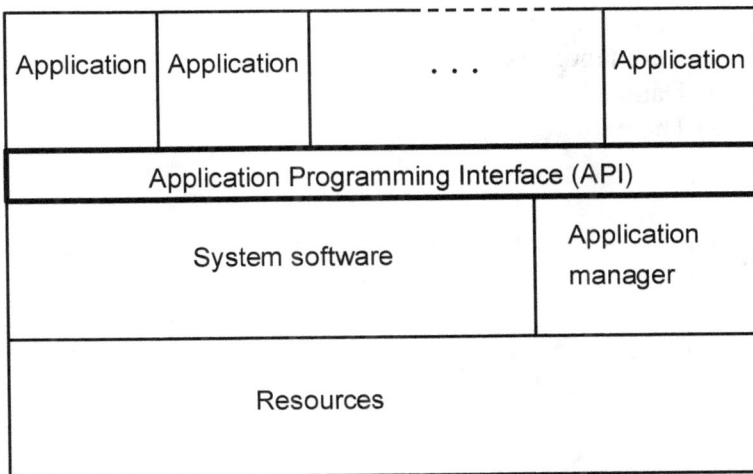

The model is actually made up of only three layers as shown in *figure 12.2*.

MHP architecture

Application	Application	. . .	Application
Application Programming Interface (API)			
System software			Application manager
Resources			

Figure 12.2

12.2.1   Resources

The resources relate to the hardware functions and therefore the drivers associated with them. A resource could also be a pure software resource. There is no restriction in the MHP model to the number of hardware resources. Hence more than one processor is possible. However, the resources must be seen by the layers above as a single entity.

12.2.2   System software and API

The system software is basically an abstraction layer. That is to say that any software above this layer is totally hardware independent. This means the applications above can be ported to new hardware with no modifications. The function of the API is to offer a clean interface between the system layer and the ones above. It defines a list of function calls into the system layer. The system layer presents an abstract model to the API, an abstract model of:

i) Streams played from different sources, and pipes for connecting them
ii) Commands and events
iii) Data records or files
iv) The hardware resources

12.2.3   Application manager

This software manages the lifecycle of all the applications. It is implementation specific, that is to say it depends on the underlying hardware.

12.2.4   Application

This layer realises the actual applications functions, eg the code necessary for the user to have some interactivity with the system.

12.3   MHP application to system software interface

Figure 12.3 shows a typical block diagram of an MHP enhanced broadcast / interactive TV profile system.

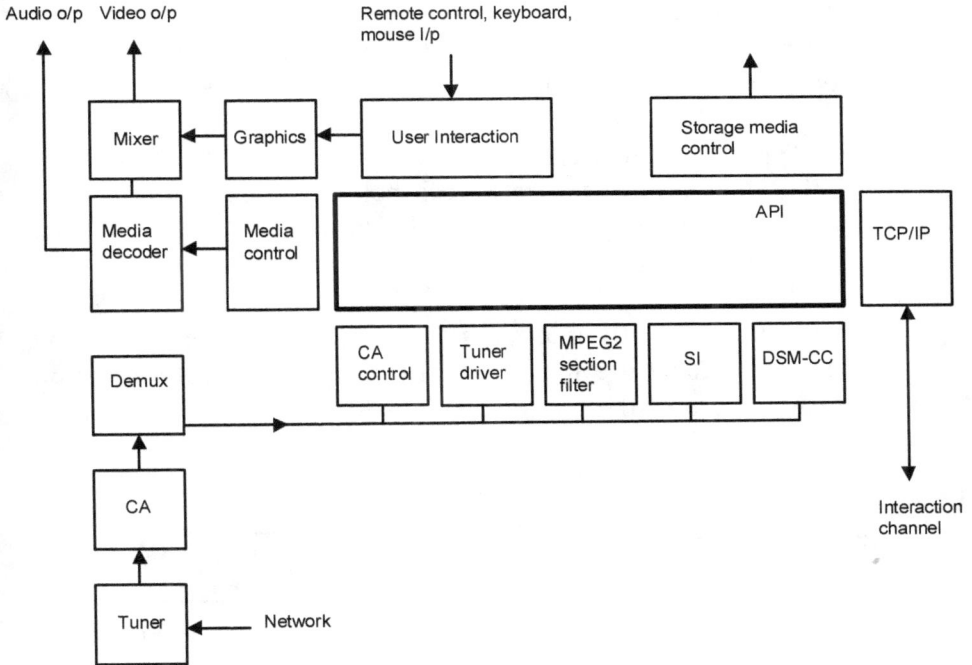

Figure 12.3

The CA control in a DVB-T system may be the CI stack software to control the interaction with the CA card. The tuner driver is there to set up the tuner parameters. The MPEG-2 section filter driver will analyse the sections of the transport stream for necessary program specific information. For example MHP dedicates a section filter to monitor the possible transmission of *do it now* events. I.e. For starting a recording or informing the viewer of a particular program start. DSM-CC is the protocol stack defining the user to user, the data carousel and the object carrousel. TCP/IP is the standard Internet protocol stack.

12.4 Plug-ins

MHP doesn't define, but allows *plug-ins* to enable older legacy software to be used within the MHP environment. The legacy system when integrated into MHP must work in exactly the same way as it did in its own legacy environment. There are two *plug-ins* supported.

i) Type A is an MHP application. Therefore it will use the MHP API and so the resources below it. This is termed an *interoperable* plug-in. Such applications may be downloaded from the network.

ii) Type B directly interfaces to the resources themselves and therefore don't use the MHP API. This will therefore remain resident in an MHP system. This is termed *an implementation specific* plug-in.

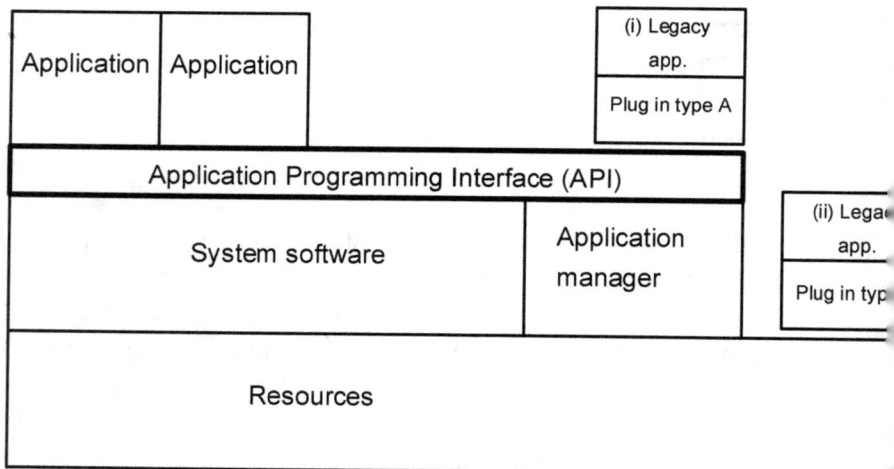

Figure 12.4

It is via this plug-in method that middlewares such as OpenTV or MHEG-5 can be modified to run over and in conjunction with an MHP environment.

12.5   Transport protocols

For MHP applications to talk to the outside world there must be some network protocols. Other DVB specifications define the network specific ones, i.e. DVB-T, DVB-S, DVB-C, and also the non specific ones (via the DSM-CC protocols). See the end of this book for the specification references.

12.6   Interaction channel protocols

As part of the MHP system software the following is the architecture for the interaction channel protocols that are accessible to MHP applications.

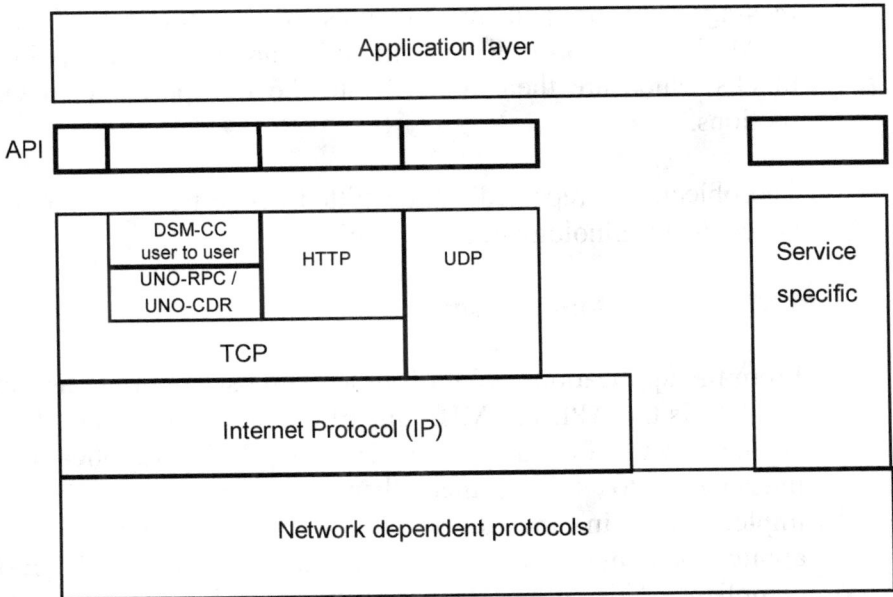

Figure 12.5

The UNO-RPC consists of the Internet Inter-ORB protocol (IIOP) as specified in CORBA/IIOP[2]. UNO-CDR is also defined in this specification. HTTP is the Hypertext transfer protocol and is defined in RFC 2616[42]. Since MHP is not network specific, the interactive channel can be via any technology such as ISDN, ATM, telephone modem etc.

12.6.1   The DSM-CC object carousel

This protocol facilitates the transmission of a structured group of objects from the broadcaster to the STB receiver. Various objects such as files and directories are inserted into the transport stream using the object carousel protocols.

This is done as follows: Objects from the object carousel are packaged up in *broadcast inter-ORB protocol* (BIOP) messages. This BIOP format allows identification of the object type and object identification through the object key. BIOP messages are grouped into modules that are transmitted in a DSM-CC data carousel. The modules are then split up into blocks which are then inserted into the transport stream via sections.

The objects are repeatedly transmitted in the transport stream hence the terminology of a *carousel*.

12.7   Applications

From the applications point of view the most important aspect of MHP is the API. For MHP one of the most important API's is the Java API, known as DVB-J. So all applications interfacing to this (other than type B plug-ins) are implemented in the Java programming language. Java applications must run the Xlet interface to be DVB-J compliant. This allows the application and the application manager to communicate bi-directionally. There can be from zero to many applications that come with a particular TV service or channel. Signalling is specified as part of the MHP specification to monitor how many applications are associated with that service along with the application life time. An application is either loaded, paused, active or destroyed at any particular point in time. Applications run on the Java *virtual machine* (VM). The applications can either be resident on the STB, or more generally will be downloaded from the broadcast channel, inside the transport stream. Incidentally it is quite feasible to write an MHEG-5 engine as an application in Java to run in an MHP environment, provided there is enough processing power to make sure it performs as it should.

12.8 The Java virtual machine (VM)

This VM (first developed by Sun Microsystems Inc.) is essentially a software CPU (although it can quite easily be implemented in silicon). It has an instruction set and manipulates various memory areas at run time. Whereas a compiler of the C language, for example, would compile to produce a machine code specifically for the processor it must run on, Java source is complied into *byte-code* or sometimes called *J-code*. This is a universal code which is interpreted by a run time interpreter. Nothing regarding the Java specification is left undefined, hence no implementation specific cases exist. Java applications are therefore truly a platform independent and universal.

The Java platform (programming language and VM) was initially developed to address the problems of building software for networked consumer devices. It was designed to support multiple host architectures and to allow secure delivery of software components. To meet these requirements, compiled code had to survive transport across networks, operate on any client (PC, MAC, or STB etc), and assure the client that it was safe to run.

The Java VM knows nothing of the Java programming language, only of a particular binary format, known as the *class file format*. A class file contains Java VM instructions (*byte-code*), ie executable code and data, as well as other ancillary information. For security reasons, the Java VM imposes strong format and structural constraints on the code in a class file. However, any language with functionality that can be expressed in terms of a valid class file can be hosted by the Java VM. Java is such an important part of interactive STB's and will become so in the future that a separate section is dedicated to give more detail regarding Java. See section 14.0 of this book.

12.9 Interoperability

Often used are the terms vertical and horizontal markets. Let's first define what these mean in the context of digital video broadcast

12.9.1 The vertical market

In a vertical market place a particular broadcaster transmits programming information in a certain well defined format. He knows his STB's will be able to decode this format. Another broadcaster also has his own standard and customers with his format STB's. This is vertical since it ties the customer in to a particular broadcaster, and to a particular STB manufacturer.

12.9.2 The horizontal market

In a horizontal market there are many STB designs giving the user the features he wants, the performance he wants at the price he is willing to pay. The broadcaster must then make sure his services can be decoded by all the different receivers on the market. The end user will expect to receive the services of all the broadcasters on the STB he decides to buy. Clearly the STB market today is a horizontal one.

12.9.3   Conformance testing

It is because of this that interoperability is vital. To ensure that each STB complies to the MHP specification the DVB conformance tests were defined. MHP265 is the DVB's conformance and interoperability tests document. The aims of interoperability are two fold:

i) All broadcasters or producers of MHP compliant application content should guarantee that their applications operate in a predictable way on any MHP compliant platform.

ii) All manufacturers of MHP compliant platforms should guarantee that all compliant applications operate in a predictable way, given an MHP compliant supporting infrastructure.

The central part of conformance testing is the API, and so many of the tests will be Java applications running on the API, therefore checking out and testing many parts of it. It is actually up to the STB manufacturers to use this *test suite* and to make sure they pass all the tests (many thousands) . the DVB have also defined mechanisms whereby conformance can be demonstrated.

12.10   Hardware requirements of MHP

Although this is not specified the typical requirements of an MHP solution are:

MHP ver 1.0:

i)   80-130 MHz CPU (not including running DVB-HTML)
ii) 16Mbit flash
iii) 32 Mbit SDRAM (including MPEG-2 decoding)
iv) Optional hard disc drive

MHP ver 1.1:

i) 150-200 MHz CPU
ii) 32 Mbit flash
iii) 32 to 64 Mbit SDRAM (including MPEG-2 decoding)
iv) 20 to 40G Byte hard disc drive

High end solutions:

i) 300+ MHz CPU
ii) 32 Mbit flash
iii) 64 to 128 Mbit SDRAM
iv) 40G Byte hard disc drive

13.0   The Java language

Java was first developed by Sun Microsystems to address the area or server client applications. It soon became popular with the advent of the World Wide Web, where it was used to add interactivity and advanced animation to web browsers. With the ability to have the applications downloaded and run on any platform it has clear and obvious applications to the STB arena. The aim of this section is simply to explain the basic architecture and some of the general terms used such as Java classes, applets and beans.

13.1   Java classes

The basic building block of a Java program is the class. In terms of syntax it is similar to a 'C' *struct*. It contains variable definitions, code performing a particular function and data. The class can be considered as a basic design such as, for example, the drawing of a window. The window (or any) class can then be invoked as many times as is necessary. Each time it is invoked it is said to be an *instance* of the class. An instance of the window class in our example. Each instance can be slightly different due to the data, thus giving the window a different look, size, colour etc. The actual function performed on the data is called the *method*. The instantiated class is termed the *object*. The data in a class are called *fields* or *variables*.

There are many predefined classes in Java. These classes are arranged in a hierarchical fashion. These classes can be put together in different ways to perform more useful tasks and actions. (The windowing type classes are actually *Jcomponent* classes and are used for building user interface systems).

13.2   Java applets

Applets are embedded Java applications (compiled class structured Java code) which need some type of viewer such as a web browser to allow the application to be realised. The class files are downloaded from the network as they're needed hence saving memory space and processing resource until it's needed. The applet knows when it is visible to the browser viewer, when it is not, and when it is no longer needed. It therefore acts relatively independently until it is no longer needed, and then cleanly finishes. The applet is not a like a full Java application in that it is restricted to the browser environment and therefore can't access files or external applications.

13.3   Java beans

A Java bean is simply a set of rules that Java objects must obey. The point of these rules is to allow applications to be built in such a way that they can very easily be reused and bolted together in *a plug and play* fashion to create the applications. Generally beans are intended to operate in a visual context, and so are put together in the construction of a graphical user interfaces. Beans, when put together after compilation, have the capability of being able to access each others methods, values and fields and so have the same capabilities as though they had been compiled together.

The *bean box* is a development environment that is supplied with Sun Microsystem's bean development kit (BDK) to allow beans to be instantiated and tested.

IV   ATSC Principles

14.0   ATSC 8-VSB system

VSB is a form of amplitude modulation (AM) which was adopted as the US standard for digital television transmission in 1996. The so called *grand alliance* recommended the 8-VSB system for digital terrestrial transmissions for the US.

14.1   ATSC implementation of the 8-VSB system

There are a number of possible variations of VSB, ie 8-VSB, Trellis 8-VSB 16-VSB etc.

The ATSC have selected the trellis 8-VSB variant. The 'trellis' refers to a method of adding redundancy to the data, such that a receiver can correct for certain types of errors. This will become clear later. The number 8 refers to the signal levels that are used to represent particular bit combination, as will also be explained later.

14.2   The ATSC trellis 8-VSB spectrum

Figure 14.1 shows the VSB system compared to the standard US NTSC spectrum. The VSB system was designed to operate over the same 6 MHz channel bandwidth as the NTSC system. It was also designed to operate without interfering with the existing NTSC analogue channels.

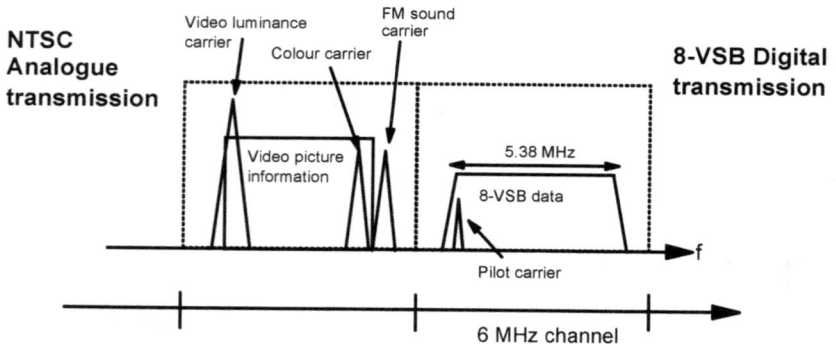

Figure 14.1

148

14.3　　　　Structure of the trellis 8-VSB data frame

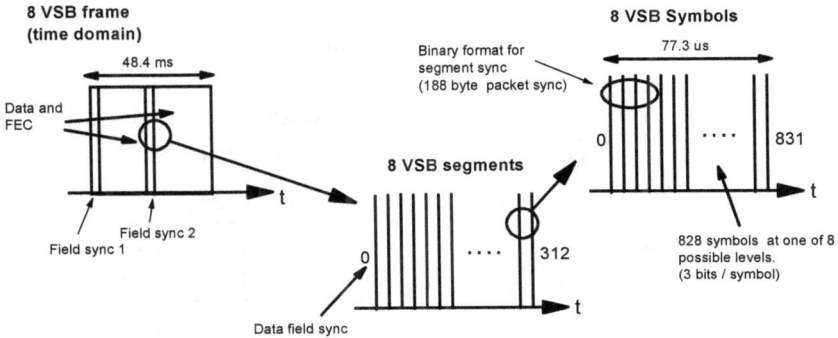

Figure 14.2

As can be seen from *figure 14.2*. The data is organised as two data fields and is transmitted over a period of 48.4 ms. As will be explained later, the frames are made up of pseudo random data sequences, with error correction bytes, data interleaving and trellis coding. The data frames are made up of 313 data segments. The first of which, in each frame, is a special synchronising signal called a *data field sync*. This is used by the equaliser in the receiver to find the start of each frame. Each data segment consists of 832 symbols, the first four of which (in each data segment) are transmitted in binary (ie at two levels only, one level representing a 0 and the other a 1). These two sync fields make up a byte which is actually used as the 208 byte transport packet sync byte. (1 sync byte + 187 data bytes + 20 Reed-Solomon bytes = 208 bytes in total). So, of the 832 symbols, the remaining 828 symbols are use to encode the remaining 207 transport packet bytes. The 828 symbols can each have any of 8 different levels. Each level is used to indicate a 3 bit sequence. For example the level -5 indicates the data sequence 011 (LSB first). The data is however trellis coded with a 2/3 trellis code, and so 3 bits are transmitted for every 2 actual or unique data bits. Hence the

149

828 symbols encode 2 x 828, or 1656 unique data bits. This constitutes the 207 bytes required.

14.4   Overall transmission sequence

Figure 14.3 shows the main functional blocks required to transmit the ATSC trellis 8-VSB programming services.

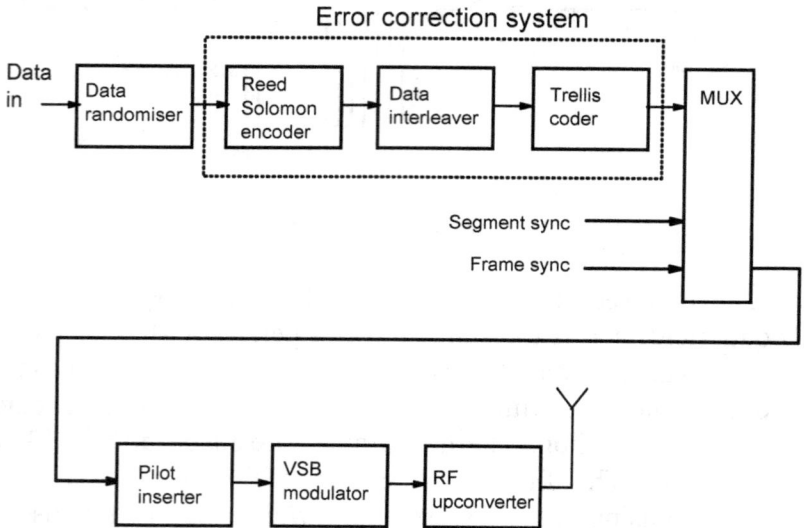

Figure 14.3

Each block will be described below, but first a few words regarding the error correction system. As has been explained in section 4.0 of this book, the channel from the transmitter to the receiver is said to be a quasi error free channel (QEF) when the number bit error ratio is of the order of 10^{-10}. Ie an acceptable number of erroneous bits is 1 in every 10^{10} bits transmitted. This is achieved by the three blocks highlighted in *figure 14.3*. The *Reed-Solomon encoder*, or more generally called an *outer coder*, adds additional bytes to the end of a block of data so that the receiver can use these to identify and correct errors introduced in the block during transmission. However only a limited number of errors can be identified and corrected. This would be a problem in the case of bursty noise effects creating many errors in a single block of data; however, the data interleaver block simply spreads out these errors, thus making the Reed-Solomon system more effective.

The final block in the error correction system is the *trellis encoder*. This is an additional error correction technique that corrects for errors, not taken into account by the previous blocks. It operates on a continuous bit stream rather than on blocks of data. A trellis is a wooden lattice like structure constructed to allow roses and various climbing plants to grow. The term 'trellis' is used in the field of convolution coding since the diagrams used to describe the behaviour of such encoders look like such a structure. The convolution encoder operates as a finite state machine, and so has states with state transitions. The diagrams of these states and their transitions are known as *trellis diagrams*. The combination of these error correcting techniques produces the QEF channel between the transmitter and the receiver that is required for digital TV data transmission.

The various blocks in *figure 14.3* will now be looked at in more detail:

14.4.1 Data in

The data input is exactly the same as that for COFDM, as explained in section I. This is therefore the MPEG 2 transport stream containing a number of standard definition digital program streams or one (possibly two) high definition program streams.

14.4.2 Data randomiser

The reason for the randomiser is the same as the energy dispersal explained for COFDM. The randomisation is however done in a different way. This is a method of removing local DC levels in the signal which would otherwise give the receiver a problem at the other end of the channel. The randomiser, randomises the data payload with the exception of the data field sync, the data segment sync and the RS parity bytes. A 16 bit maximum length pseudo random binary sequence (PRBS) is started at the beginning of the data field. The PRBS is generated in a 16 bit shift register with 9 output taps. These taps are individually XORed with their corresponding input data bits as shown in *figure 14.4*:

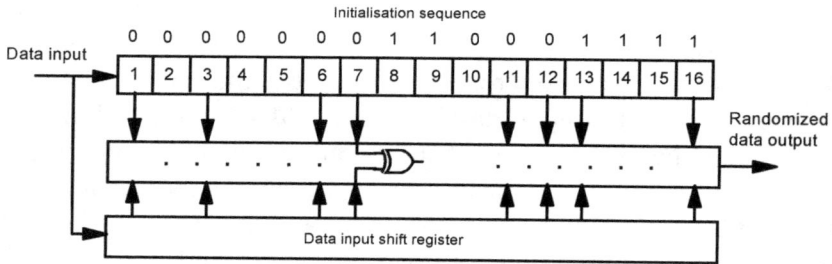

Figure 14.4

The initialisation sequence is:

$$X^8 + X^9 + X^{13} + X^{14} + X^{15} + X^{16},$$

and the randomiser polynomial is:

$$1 + X + X^3 + X^6 + X^7 + X^{11} + X^{12} + X^{13} + X^{16}.$$

14.4.3   Reed-Solomon encoder

This is applied to allow a receiver to work out which bytes have been corrupted during the transmission, and to correct for them. It can correct up to 10 corrupted bytes, and adds 20 additional bytes to the end of the 188 byte scrambled transport packet for this purpose. These 20 additional bytes are known as Reed-Solomon bytes. This coding technique is the same as is used for the COFDM outer coding, with some important differences in parameters used. The 8-VSB system uses Reed-Solomon (RS207, 187) as opposed to Reed-Solomon (RS204, 188) in the case of COFDM. The impact of this will be explained in section 15.0 which compares the COFDM and the 8-VSB systems.

Reed-Solomon coding is mathematically very complicated and so will not be explained here in any detail. The aim is simply to give an overview of the coding technique. The coding is a block level code. That is to say that it operates over a block of data. The block of data must therefore be constructed prior to the code operation being performed. This puts an overhead on the system in terms of memory and the need to block synchronise. The coding adds 20 additional bytes to the 188 byte transport stream packets, making a final transport stream packet size of 207 bytes. This error correction algorithm, as applied to a transport stream, is characterised by the three numbers n, k and l. Where n is the number of bytes in the final transport stream, k is the number of bytes of the original transport stream, and t is the number of bytes that can be corrected. So:

$n = 207$ (final transport packet length)
$k = 187$ (Original transport packet length)
$t = 10$ (Number of correctable bytes)

The Reed-Solomon code effectively specifies a polynomial by generating a large number of points. The Reed-Solomon code detects errors within these points much as the human eye can detect errors in what should be a smooth curve. The Reed-Solomon code can detect and correct up to $(n-k)/2$ errors. The field generator polynomial used is:

$$P(X) = X^8 + X^4 + X^3 + X^2 + 1$$

with the code generator polynomial:

$$G(X) = (X + \alpha^0)(X + \alpha^1) \ldots \ldots (X + \alpha^{19})$$

14.4.4　　　　Convolution interleaving

This is equivalent to the Outer interleaving or *Forney convolution interleaving* explained for COFDM. It is performed to basically spread out the errors (in time) and so make the Reed-Solomon coding more effective. As stated above, the Reed-Solomon coding can correct up to 10 bytes in a transport stream packet. Clearly if a burst error condition occurs, ie a burst of energy from some noise source, then more than 10 bytes within the same packet could become corrupted. The convolution interleaving effectively takes these errors and spreads them out over a number of packets, thus allowing the Reed-Solomon to be more effective.

Figure 14.5 shows the transmitter architecture. Data is input from the Reed-Solomon outer coder and output to the trellis coder.

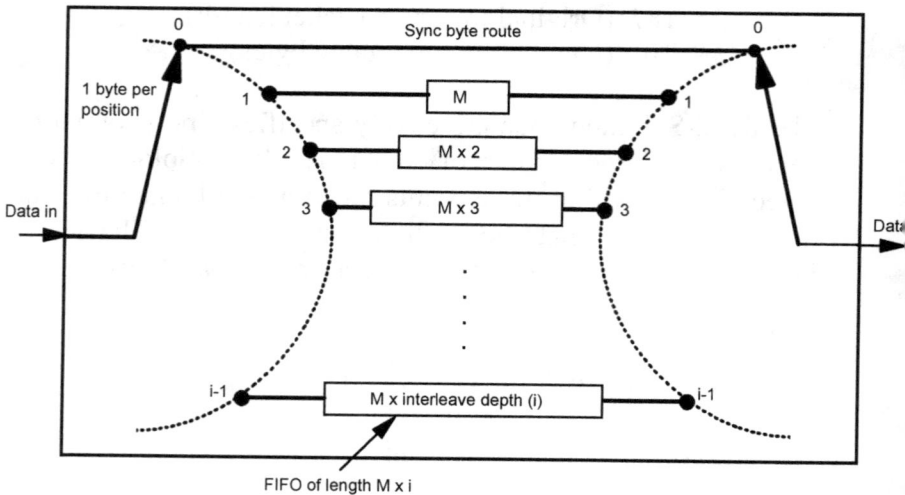

Figure 14.5

The following are the dimensions of the interleaver for the 8-VSB system:

i = 52 (Number of branches, therefore shift registers)
L = 208 (Length of the packet to be protected)
M = (L/i) x j (Size of the FIFO's in bytes)
j (An index that ranges from 0 to i-1)

Therefore there are 52 individual branches, with the largest FIFO being of length 204 bytes. Input bytes will therefore be delayed by 4, 8, 16, ... 204 bytes, depending on the byte index. Note that the sync byte always passes through directly and therefore experiences no delay.

Figure 14.6 shows the architecture at the receiver side.

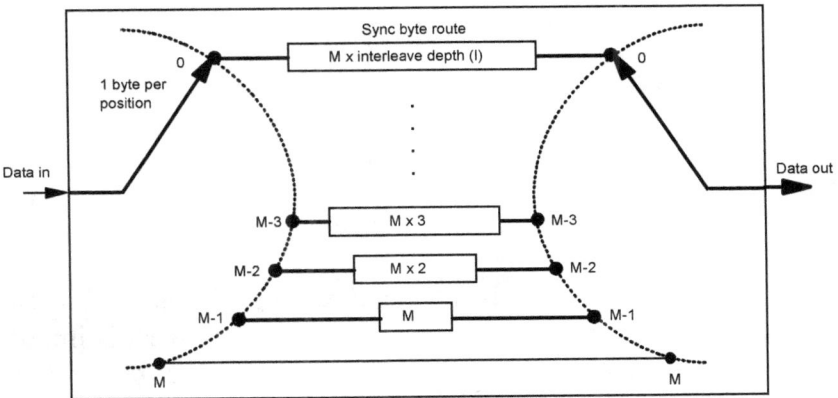

Figure 14.6

Data bytes with j index of 51 were delayed by 204 bytes at the transmitter, and so at the receiver the j = 51 bytes experience no delay, while the j = 0 bytes experience the maximum delay of 204 bytes. The original data stream is therefore reconstituted.

14.4.5   Trellis coding

This is in fact a convolution coding and is the same as the inner (or sometimes called Viterbi) coding used with the COFDM system. The main purpose of the trellis coding is to add redundant data into the bit stream to allow a receiver to correct for non impulsive errors. The ATSC system in fact uses twelve identical trellis encoders acting on twelve groups of interleaved symbols. The concept of trellis coding can be explained as follows:

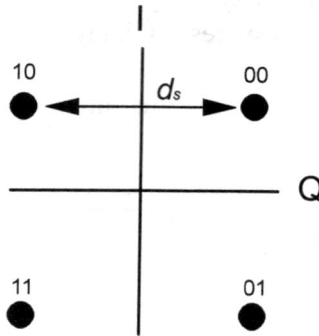

Figure 14.7

Trellis coding effectively identifies the distances d_s between constellation points to reduce the error rate. What is important in trellis coding is the transition from one bit pattern to another, this transition is only allowed between certain bit patterns. When a transition occurs the system is considered to be in a specific *state*. It is the representation of these states that gives rise to the trellis like diagrams that give the technique its name. Only certain transitions are allowed. If a transition occurs violating this, the system is said to be in an error state.

Lets look at the way a receiver works to better understand this before we go any further. As has already been said, the symbols, which are simply pairs (or more) of bits together, when suitable mapped onto I and Q axis will look something as shown in *figure 14.8*. Let us assume for now that there is no convolution encoder.

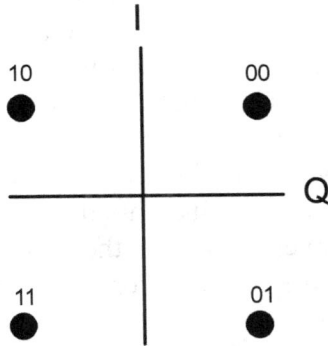

Figure 14.8

This therefore is the constellation that the transmitter system is generating. However in heavy noise conditions this constellation will become degraded and may look more like as shown in *figure 14.9*, as seen at a receiver.

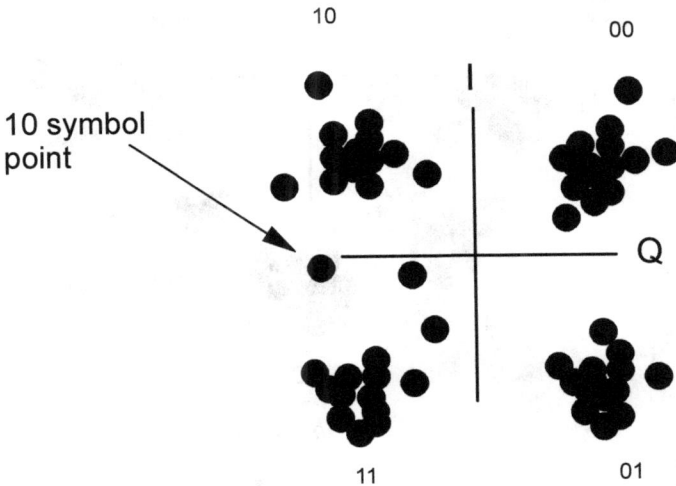

Figure 14.9

Most of the symbols will appear in the neighbourhood of the theoretical positions, as was transmitted. However there is a probability that some will be farther away. This being the case, the receiver will have a problem identifying the correct locations of the points, and so will make some errors. These errors are very much related to the distances the points are away from each other. *Figure 14.9* shows, for example a point that the receiver may interpret as being 11 instead of 10. (Clearly there is less likelihood of an error occurring that causes the receiver to identify the point as 01, as this is further away). What the convolution encoder achieves with, for example an *n = 2* encoder, might be as shown in *figure 14.10* at the receiver side in heavy noise conditions:

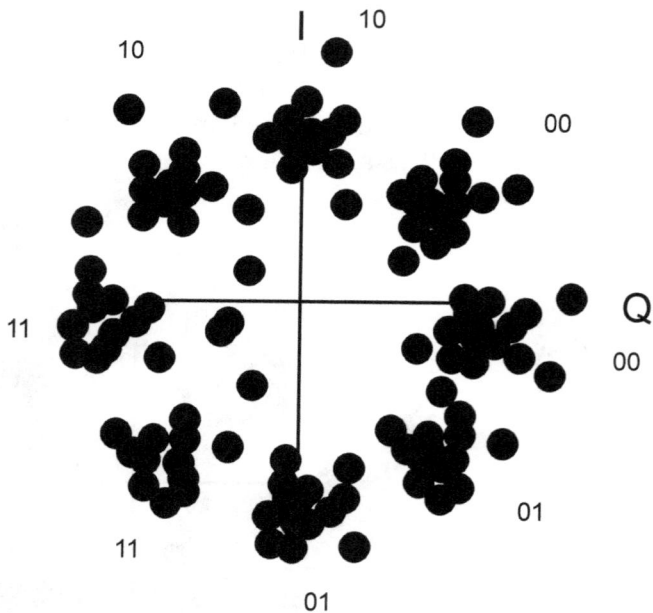

Figure 14.10

The transmitter now has two outputs. These are mapped onto two constellations and offset in phase. Therefore *figure 14.10* shows what a receiver may see in heavy noise conditions for a QPSK constellation.

Now lets take a symbol that the receiver may have a problem with. For clarity we will label the constellation points as A, B, C etc.

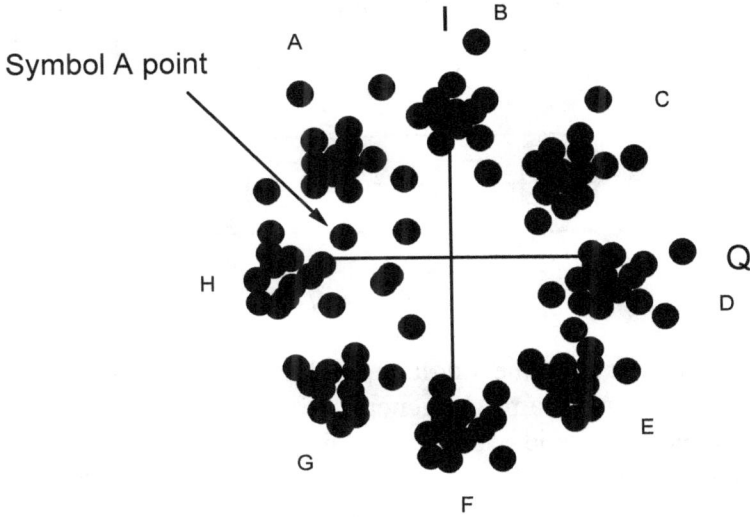

Figure 14.11

The symbol indicated in *figure 14.11* may be interpreted by the receiver as being an H point, or if its position were even worse, a G point. However the convolution encoder will not allow certain symbol transitions. If, for example the previous point to be transmitted was an A point, then, for example, it may not be possible to have the next point appear in position H or position G. Therefore the receiver would know that this point is in error. It would then work out which point is the most probable for this erroneous symbol, and therefore correct it.

14.4.5.1   Operation of trellis coding

The technique introduces redundancy by increasing the number of symbols. Section 4.4 introduced the COFDM convolution encoder, the general architecture of which is shown in *figure 14.12*:

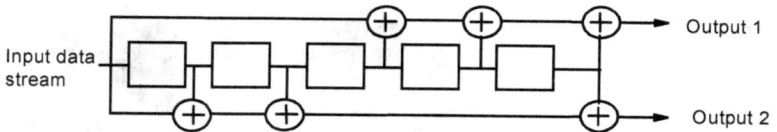

Figure 14.12

In this example the single input data stream is used to generate two output streams. In general this type of architecture can be increased to add more redundancy. In fact the ATSC system uses a similar scheme as shown in *figure 14.12*, but with the input bit stream also being output, as shown in *figure 14.13*:

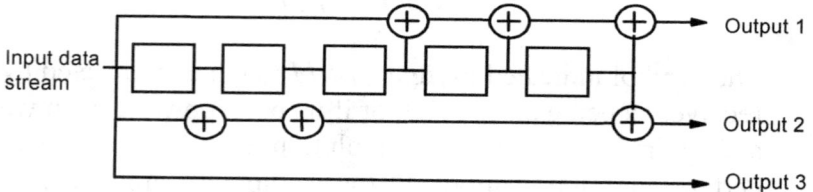

Figure 14.13

Note that *figure 14.13* is only a representation, the adder units are not shown in their true positions, and in fact there are only two shift register stages used in the convolution encoder for the ATSC system, hence 4 states. The number of adder units is described by a particular polynomial.

To better understand the system lets look at the $n = 3$ convolution encoder in more detail. Further, lets assume that it is a very simple one with only 2 shift register stages and two modulo two adders (see the glossary if you are unfamiliar with modulo two addition), as shown in *figure 14.14*. (The shift register contents are set to zero at the start).

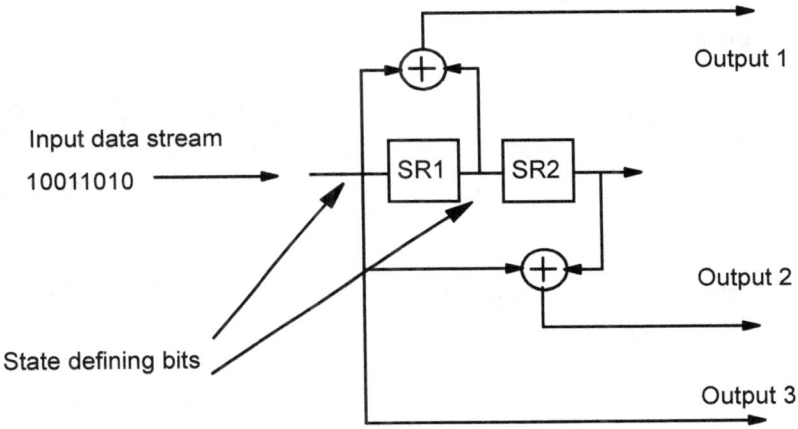

Figure 14.14

Lets follow the bit pattern from input to output on each clock edge. *Table 14.1* shows the situation after the first clock. Ie when the first 0 of the input data has been clocked to the output of SR1.

Input	SR1 output	SR2 Output	State	Output 1	Output 2	Output 3
0	0	0	S0	0	0	0
1	0	0	S2	1	1	1
0	1	0	S1	1	0	0
1	0	1	S2	1	0	1
1	1	0	S3	0	1	1
0	1	1	S1	1	1	0
0	0	1	S0	0	1	0
1	0	0	S2	1	1	1

Table 14.1

The bits entering SR1 and SR2 are assigned a name or more conventionally a state. Ie 00 is state S0, 01 is state S1, 10 is S2 and 11 is S3. It can be seen that when the encoder is in state S0 it can only move into either state S2 or remain in state S0. The same can be said for every state. There are only 4 possible states allowed in this example, simply due to the 2 levels of history. Clearly with 3 levels of history there would be 8 possible states.

It is important to not confuse states and levels. The number of states is a function of the number of shift registers, ie the amount of history associated with the convolution encoder. When talking about an 8-VSB system, the 8 refers to the number of levels. These levels are signal levels, one level for each bit pattern (or symbol) leaving the convolution encoder. 3 bits per symbol, hence 8 levels.

If *table 14.1* is carried on with a random bit stream as input it will soon become apparent that states can only transition to certain other states. Ie state S0 can only stay in state S0 or move to state S2. S1 can either move to S0 or to S2. S2 can either move to state S1 or S3. S3 can either stay in S3 or move to S1. *Figure 14.15* shows this graphically.

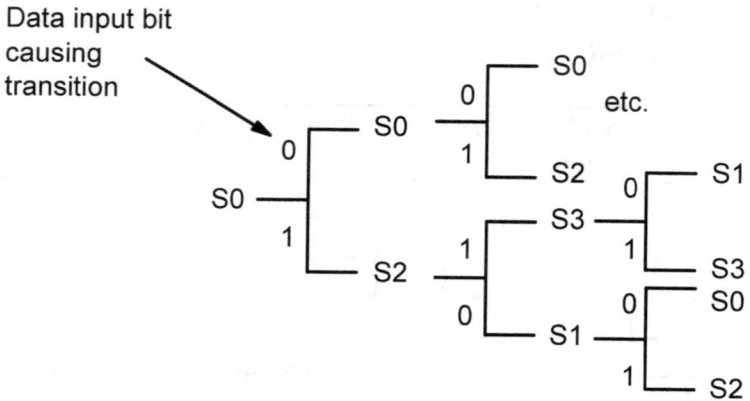

Figure 14.15

The data input bit is the bit that causes the state transition. This bit is shown in *figure 14.15*.

Another way of representing this is in the form of a trellis diagram, as shown in *figure 14.16*, this shows the input data bit having caused the transition in brackets.

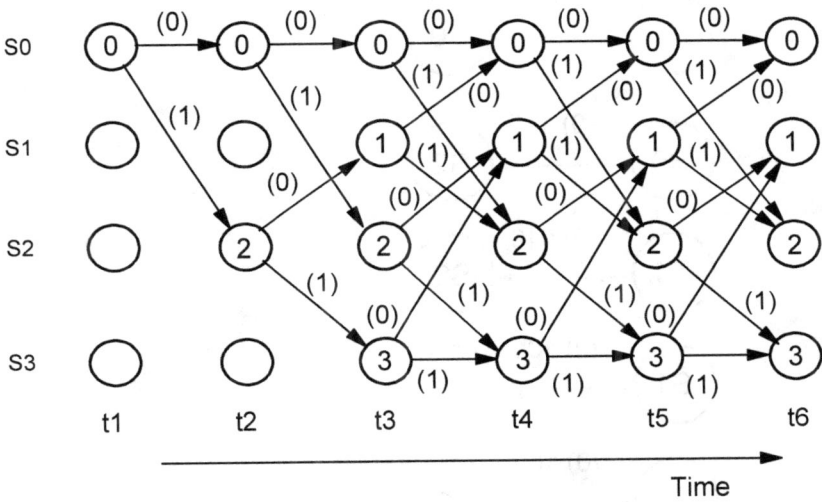

Figure 14.16

The states are shown with increasing number from top to bottom on the trellis diagram, with the allowed movements depending on whether the next convolution encoder input bit is a 1 or a 0. *Figure 14.16* shows the progress of the state encoder with time. As can be seen; at the beginning it can either remain in S0 or move to S2 depending on the input data bit value. (ie a 0 will cause it to remain in S0 and a 1 to move to S2). From time t2 to t3 the possibilities are drawn in for both cases of the previous input bit value and so on as time progresses.

In fact looking at *table 14.1*, the outputs 1, 2 and 3 become as shown in *figure 14.17* as the states move to their allowed next states depending on the input bit shown in *figure 14.17* in brackets:

Input bit causes state transition and outputs 1,2, and 3

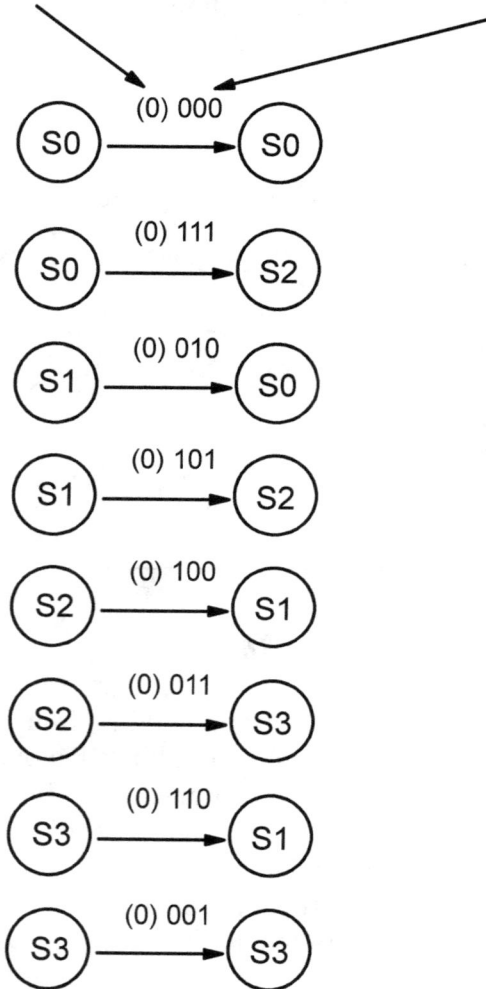

Figure 14.17

It can therefore be seen that the data input of *table 14.1* gives an output bit stream of:

000 111 100 101 011 110 etc. with the states moving from: S0 to S2 to S1 to S2 to S3 to S1, this is shown as bold lines on the trellis diagram of *figure 14.18*.

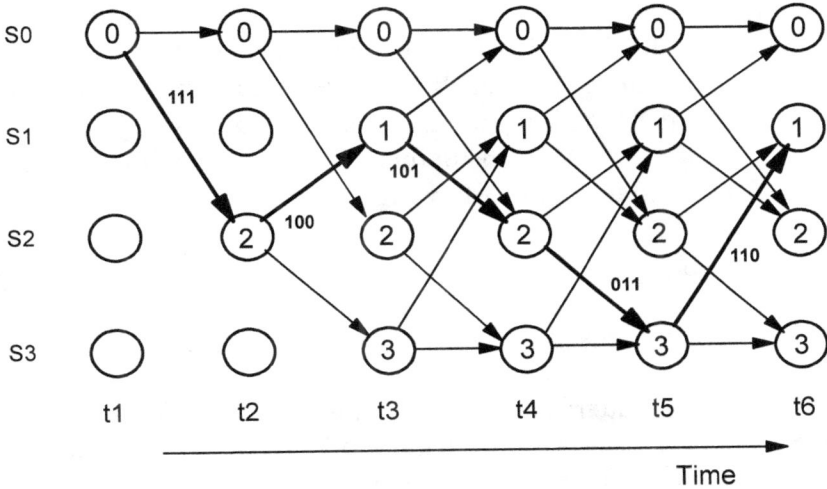

Figure 14.18

When moving from one state to another the bit pattern will have a different distance d_h from the input bit stream than another. The smaller this distance the greater the probability that the symbol belongs in the particular state. For 8-VSB this distance is described by the *Hamming distance*. A small diversion will be made now to describe this and will then be calculated for the example being considered. Incidentally, these groups of three convolution encoder output bits are now termed as *symbols*.

14.4.5.2   Hamming distance

The hamming distance is simply a description of how far away a particular bit pattern is from another. Ie the pattern 111 is a long way from 000 since if 111 were transmitted but 000 were received, this would need 3 errors to achieve, which is less probable than say 1 error. It is therefore said that the distance between the two bit patterns is large. To quantify this, the Hamming distance (d_h) is defined. The Hamming distance can be calculated using the following formula:

$d_h = W(S_i \oplus S_j)$.

The way it works is as follows:

The two bit patterns (or symbols) are modulo 2 added, lets say $S_i = 111$ and $S_j = 000$. This would result in 111. W is a function that returns the number of 1's in this result, ie 3. Therefore $d_h = 3$ in this example. If $S_i = 111$, and $S_j = 110$ then $d_h = 1$.

Lets now get back to the convolution encoder operation, and in fact look at the convolution decoder, and how the Viterbi algorithm can correct for errors.

14.4.5.3   Viterbi algorithm

Lets look at the case where the following bit pattern is transmitted, ie:

111 100 101 011 110 etc.

as in the example of *table 14.1*. However lets assume that the received bit pattern is:

111 100 111 011 110 etc Notice one corrupted bit.

First the hamming distance is calculated between the 3 bit words received, and all the possible outputs of all the state transitions allowed. This is done during each time period, ie in time period t1, the only transition is 111. The hamming distance between this and the data input is clearly 0. In time period t2 there are four possible transitions.

1) State 0 can remain in state 0
2) State 0 can go to state 2
3) State 2 can go to state 1
4) State 2 can go to state 3

The hamming distance in case (1) is 2 since in this time period the input data stream is 100. S0 to S0 produces a 000 bit stream, hence the difference between these is 1. For case (2) the difference between 100 and 111 is 2. For case (3) the difference between 100 and 100 is 0. For case (4) the difference between 100 and 011 is 3. This is simply done for every transition of the trellis diagram as sown in *figure 14.19*:

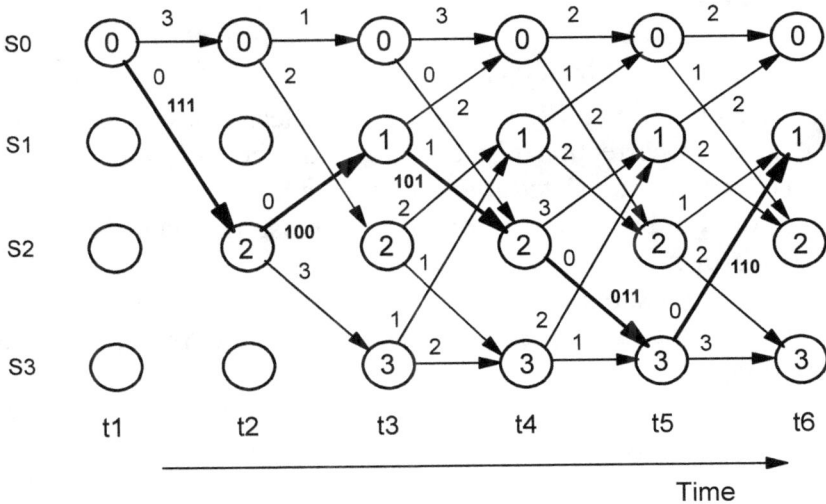

Figure 14.19

167

It is clear that an input bit stream to a convolution decoder can only be allowed to transition between certain states only. So when a bit pattern is received it will have a choice of one of two new states to go to. It is the choice of which state that should be entered that is the purpose of the Viterbi algorithm. The diagram
of *figure 14.19* shows that the original bit pattern is reconstituted by selecting the route through the trellis diagram with the lowest hamming distance. Hence the original bit pattern;

111 100 101 011 110 is reconstituted with a total hamming distance of 1.

The Viterbi algorithm calculates the distance up to each state at a given time and deletes the path with the greater distance. For example at time t4 in *figure 14.19* state 2 has two inputs. The accumulated hamming distance of the path leading from state 0 is 4, and of the one coming from state 1 is 1. Hence the path leading up to state 2 at time t4 is deleted.

The number of convolution encoder stages chosen, and therefore the number of trellis states, clearly has an impact on the error correction capabilities of the convolution encoder. This can be translated into a gain measurement. This can then be used to define how far above the noise floor a system must be to operate at a certain acceptable error rate. *Figure 14.20* gives an example of the improvement of a trellis system compared to a simple QPSK modulated system with no trellis coding.
The graph of *figure 14.20* shows the trend as trellis states vary from 4, to 8 to 16 to 32 and to 64 states.

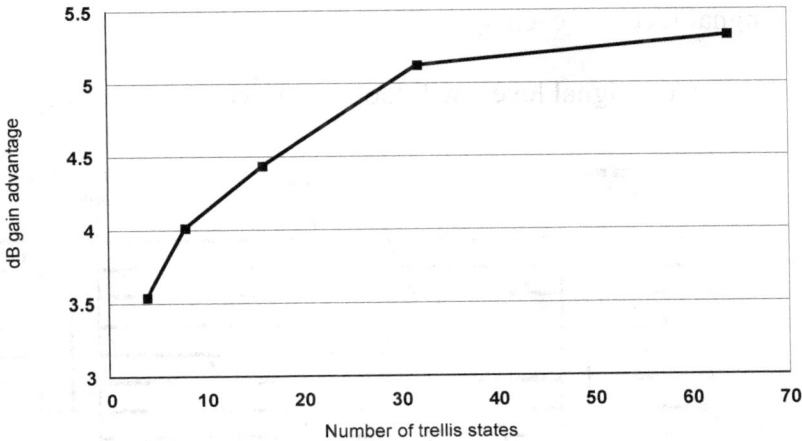

Figure 14.20

Clearly, as the number of states increases, the gain advantage has diminishing returns.

14.5   VSB modulation

The remaining important block in the transmission sequence is the VSB modulation block. This is simply a matter of taking the output of the twelve trellis coders and mapping the symbols generated onto 8 signal levels as shown in *table 14.2*.

Output 3	Output 2	Output 1	R
0	0	0	-7
0	0	1	-5
0	1	0	-3
0	1	1	-1
1	0	0	1
1	0	1	3
1	1	0	5
1	1	1	7

Table 14.2

Output 1, 2 and 3 therefore define the symbol and R is the signal level for each symbol.

Hence the signal levels will look as shown in *figure 14.21*.

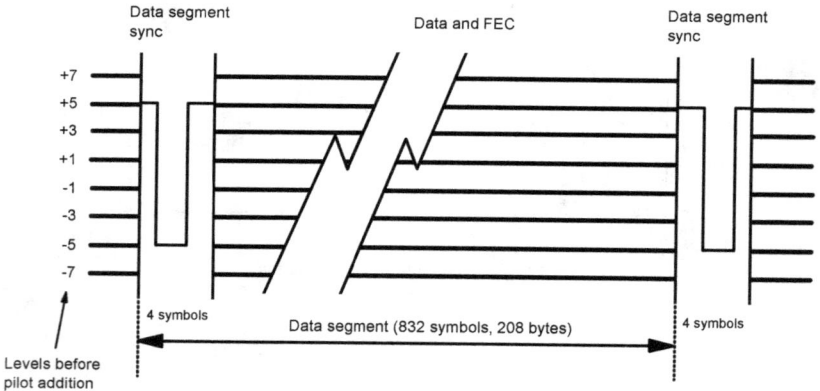

Figure 14.21

The small DC value of 1.25 is added to every symbol (data and sync) after the bit to symbol mapping in order to create a small pilot carrier.

14.6   8-VSB useful bit rate

The useful data rate can be calculated from the following equation:

Useful data bit rate $= \dfrac{D \, x \, S \, x \, M \, x \, C \, x \, F}{T} x \, B \, x \, Fs$ bits per second

$$= \frac{313 \, x \, 832 \, x \, 3 \, x \, 2/3 \, x \, 2}{48.4 \, ms} x \frac{188}{208} x \frac{312}{313}$$

$$= 19.39 \text{ Mbits per second}$$

Where:

D is the number of data segments (fixed at 313)
S is the number of symbols (fixed at 832)
M is the number of bits per symbol (fixed at 3)
C is the convolution (trellis) coding rate (fixed at 2/3)
F is the number of frames in the complete 8-VSB data frame (fixed at 2)
B is the efficiency of the RS block code (188/208).
Fs is the overhead due to the data field sync pulses
T is the duration of the complete 8-VSB data frame (48.4 ms)

Note that there is no need to take into account any correction for the segment sync pulses (You may expect to add the following factor: 828/832). This is because, as was explained earlier, these two, 4 bit signals also act as the 188 transport stream packet sync byte.

V DVB-T COFDM Vs ATSC 8-VSB

15.0   DVB-T COFDM & ATSC 8-VSB compared

The first in depth performance comparisons of the DVB-T and the ATSC systems was performed as part of the evaluation done in Australia to decide which system should be adopted for their digital TV transmissions (The DVB-T system was chosen). However since then many trials around the world have been conducted.

15.1   Main system differences

The main system differences are summarised here.

Parameter	DVB-T	ATSC
No. of carriers	1705 or 6817	1
Modulation types	QPSK, 16-QAM, 64-QAM	8-VSB
Hierarchy coding	Yes	No
Outer coding (RS)	8 byte errors	10 byte errors
Inner coding	Convolution (64 state)	Convolution (4 state)
FEC coding rate	1/2,2/3,3/4,5/6,7/8	2/3
Energy dispersal	15 bit shift register with 3 taps	16 bit shift register with 9 taps
Bit rate variations	4.98 to 31.7 Mb/s @ 8MHz	19.4 Mb/s @ 6 MHz (fixed)
Bandwidth	6/7/8 MHz	6/7/8 MHz
C/N in AWGN (Theoretical)	16.5 dB	14.9 dB
Sensitivity to transmitter / translator linearity	Low	High
Geographic transmission type	SFN and MFN	MFN

Table 15.1

The following points should be noted when considering *table 15.1*:

15.2 Modulation types

The various modulation types are shown here for the DVB-T system since it is generally possible to select any of these as required. The 'grand alliance' have defined 8-VSB for terrestrial and 16-VSB for cable transmissions. Unique receivers would generally need to be built for these requirements. In fact VSB based systems have been defined as 2-VSB, 4-VSB, 8-VSB, 16-VSB and 8T-VSB. Note the 8T-VSB here signifies the terrestrial variant, with a trellis code rate of 2/3. The other options are not trellis coded.

15.3 Hierarchy coding

The DVB-T system has hierarchy modes, this allows the simultaneous transmission of the same or different programming data for the following reasons:

1) Transmission of the same programs at HD and SD with greater error recovery. This allows for poor areas of reception to pick up a SD picture if the HD signal is inadequate.

2) The transmission of a program with different resolutions and characteristics to allow reception by different types of receivers, ie low cost, low resolution, mobile etc.

3) The transmission of other data related to the programs being transmitted.

See section 5.7 of this book on COFDM parameters for more information.

15.4 Outer coding

The outer coding in *table 15.1* shows the number of bytes that can be corrected by the two systems per transport packet. The Reed-Solomon coding used for the ATSC system is RS(207, 187), the one used for the DVB-T system is RS(204,188). This means that for impulsive noise sources the ATSC system is more effective. These noise sources can by such things as motorcycles, cars, industrial machinery, and domestic appliances. Although these items should be screened for such high energy emissions.

15.5 Bandwidth

It is possible to implement both systems at various channel bandwidths, however, the DVB-T system was generally designed to be implemented within a bandwidth of 7 or 8 MHz since this is what is necessary to allow the 20 MB/s bit rate with good error correction. This is a requirement for the transmission of HDTV, at maximum definition (1080 lines by 1920 pixels). Note however that a lower resolution of 720 lines by 1280 pixels is also a popular format but requiring far less bandwidth. The 8-VSB system is more efficient in terms of bandwidth since it can operate in a 6 MHz channel with comparable error correction
(ignoring performance in high echo environments), and achieve a 19.4 Mb/s bit rate.

15.6   Echo correction (static)

An important criteria to consider when implementing any broadcast system is the susceptibility to ghosting or echoes. This has been very carefully thought out, and a correction system implemented with the DVB-T system. Echoes can be caused from, in the main, signals bouncing off buildings, mountains (and walls in the case of a mobile receiver or portable TV), thus presenting two signals to the receiver, one delayed with respect to the other. Another cause is when a multiple transmitter system is implemented. Hence the same signal can be received from two different transmitters. The one further away will therefore be effectively producing a weaker echo signal.

The graphs in *figures 15.1* and *15.2* show the C/N levels at which the two systems will operate given a particular echo level. The first graph shows this with echoes of length 4.2 µs, the second with echoes of 7.5 µs. Both graphs were plotted with the DVB-T COFDM system set to 64-QAM modulation, 2/3 FEC code rate and 1/8 guard interval.

8-VSB and COFDM C/N threshold vs echo level (4.2 uS post echo)

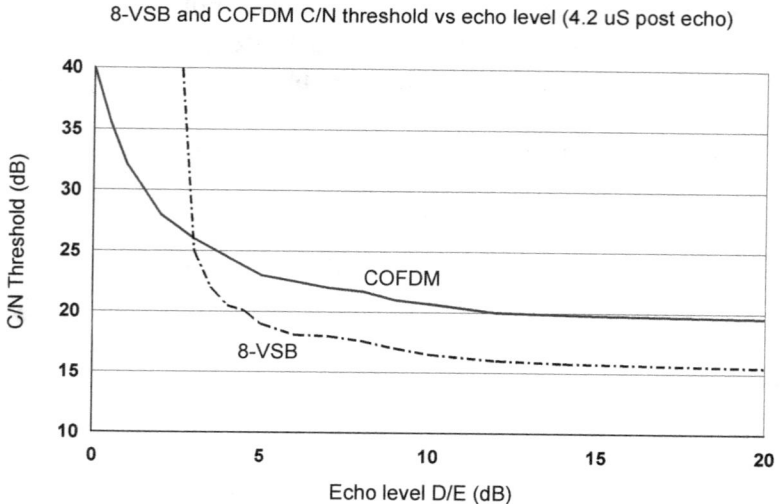

Figure 15.1

8-VSB and COFDM C/N threshold vs echo level (7.5 uS post echo)

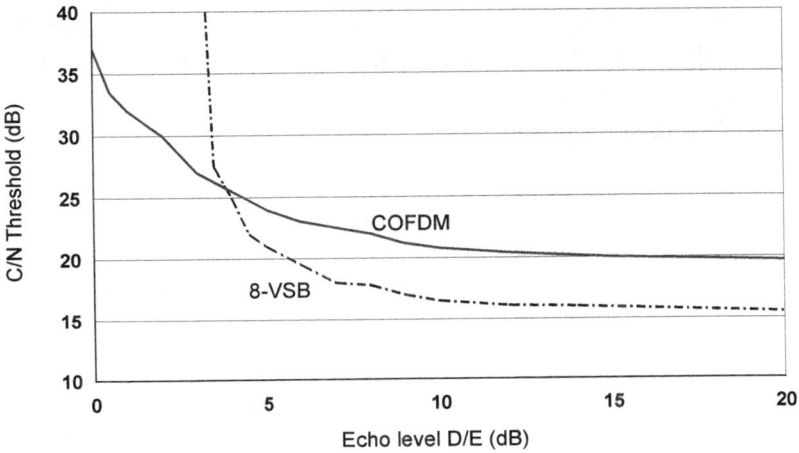

Figure 15.2

Since the ATSC system was never initially designed to operate under echo conditions it's not surprising that the DVB system is far more robust in this area. Some 8-VSB receiver chip manufactures do claim however that they have their own algorithms that allow the ATSC system to compete. The diagrams above show a better performance for the DVB-T system for high amplitude echoes and better performance by the COFDM system for small amplitude echoes.

15.7 Echo correction (Doppler)

Another source of echoes is in the case of mobile receiver reception. A mobile receiver that is moving will produce a frequency shift in the received signal. This is the same effect as a train going past a stationary observer. If the train whistle is activated, the pitch is at a high level as the train comes towards the observer, since the audio compression wave is effectively compressed and therefore at a higher frequency. As the train goes past, the audio wave is stretched out and so the frequency is reduced. The same effect occurs for electromagnetic waves such as light and VHF etc. This effect can actually also be produced from an echo signal bouncing off a moving object such as an aeroplane, car of even a leaf on a tree moving in the breeze. *Figure 15.3* shows the results obtained by the Australian test under such conditions. Note that this graph is just an indication of the nature of the performance.

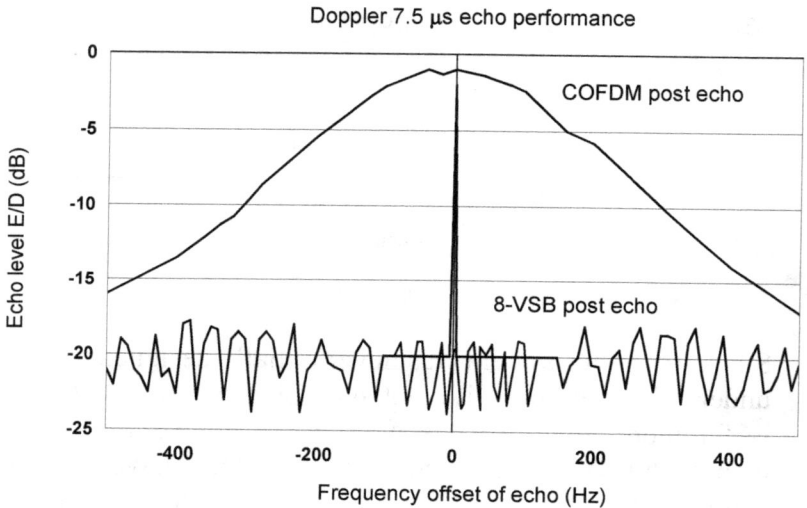

Figure 15.3

As can be seen the Doppler single echo performance is markedly different between the two systems. For example at a frequency offset of 100 Hz, the DVB-T system can perform with an echo level of around -3dB, but the ATSC system will only perform with an echo level down around 18dB.

15.8   C/N in AWGN

Additive white Gaussian noise (AWGN) is a method of simulating real world noise conditions. This parameter is therefore important when considering the range of reception and the power necessary to achieve this.

15.9 Geographic transmission type

The DVB-T system was designed to offer the capability to broadcast within a single frequency network (SFN). (Clearly also MFN's are also supported). The 8K system allows for echo correction to be performed up to 224 µs. This corresponds to an echo path difference of 67 km.

15.10 Conclusion

There are in practice many criteria that must be taken into account by counties making decisions on which system to use for their digital services broadcasts. Some of these not mentioned in this book are the limitations they may have on such things as: Geographical location, channel bandwidth allocation, economic factors and political factors. However from a purely technical view point the following generalised comments can be made regarding a choice between the DVB-T COFDM system and the ATSC trellis 8-VSB system:

- The DVB-T system is a must for broadcasters interested in reception by mobile and indoor receivers.
- The ATSC system is capable of transmission of *maximum* resolution HDTV within a 6 MHz channel.
- The performance of the ATSC system is much more related to the particular receiver system used
- The DVB-T system is suitable for SFN's, the ATSC system is not.
- The DVB-T system is far more tolerant to echo conditions. (particularly large echo amplitudes)
- The ATSC system needs less transmitter power for equal broadcast coverage area.
- The ATSC system is better at coping with short impulsive noise sources.
- Interference into current analogue channels may be slightly less with the ATSC system.

- The DVB-T system has hierarchy modes to address areas of marginal reception capability.
- The DVB-T system has more options and is more flexible to cope with future broadcast requirements.

The debate between these two main players however still continues. For example here is a short history:

Early in 1999

The US department of defence reported on tests that they had been conducting resulting in a Pentagon delegation urging the Federal Communications Commission (FCC) to revise the U.S. digital TV standard to include a new transmission option. Concerns were in the ability of the United States' existing DTV system to broadcast to the small portable receivers that consumers rely on in times of national emergency. Delegation members also made clear that they believed it would be best to build all TV equipment to one standard.

October '99

The Baltimore based Sinclair Broadcasting group (SBG) announced it had data is was about to release that would be very damming to the 8-VSB system. Further they announced that they would file a petition to the FCC to lobby for broadcasters to have the choice of transmission system. SBG further stated that in their opinion the European DTV standard, would be better than 8VSB for transmitting to inside set-top-box antennas and for portable use; the very sort of receivers that consumers would need to receive emergency information from the government in case of emergency situations

December '99

The FCC commissioner reported on tests performed by the department of defence, stating that they had serious concerned regarding the transmission to small portable receivers.

January '00

The DVB (European standards body) stated that the modifications to the VSB chip sets to address the reception issues complained of above were failing to solve the problems. The Mexican authorities, who were in the process of selecting a transmission system for their digital broadcasts, were shown a copy of the FCC's own confidential report on the subject. This report apparently said that the SBG comparative tests of 8-VSB and COFDM were 'badly biased' and represented an 'unsound and unfair attempt to discredit the ATSC standard'. They indicated that next generation chip sets would solve the echo problems.

At around the same time Motorola, in collaboration with the Sarnoff Corporation claimed to have solved the reception issues with a revolutionary signal processing architecture. However, their own web site posted data regarding field test with their latest demodulator chip showing the claims to be untrue.

January '00

Zenith carried out some comparison demonstrations or their own at the consumer electronics show in Las Vegas and in Washington. They were claiming the better performance of the up-to-date technology being used for 8-VSB transmissions. SBG were invited to these demonstrations, and stated that they considered that Zenith were making a fundamental mistake in their comparison tests by using only simulated multi path signals instead of true live signals.

February '00

The FCC rejected the SBG request for a choice in the transmission system to use for digital transmissions in their standard. The FCC simply stated that the shortcomings were in early equipment and the problems were being resolved.

February '00

Zenith re-affirmed its support for the ATSC 8-VSB system for digital transmission.

April '00

Microsoft Corporation came down on the side of the COFDM system, criticising the 8-VSB system for poor coverage and hampering emerging businesses from delivering digital services to mobile appliances. They expressed concern on the ability of the 8-VSB system to reach the same coverage as the current ATSC transmissions. This claim was challenged by the Envisioneering group of New York who, through their own testing in Manhattan, claimed that the digital signals could be received in more places than the analogue NTSC signals.

July '00

A congressional hearing in the US sees a Sinclair Broadcast Group Engineer set up and demonstrate a DVB-T TV on the witness table in front of the members. It was receiving broadcasts at 19.77 Mbits /s from a nearby DTV station with only a simple omni-directional 12" 'bow tie' antenna. The data rate being above the maximum ATSC data rate of 19.39 Mbits / s. Zenith and the consumer electronics association meanwhile conducted their own demonstrations of the ATSC system. These demonstrations used two carefully aimed directional antennas (hidden from view on window sills). When questioned about the positioning of the antennas, it was stated that there would be a potential for failure if they were moved.

At the writing of this book only three countries have adopted the ATSC system (Argentina and Taiwan having changed their minds in favour of DVB-T). Over twelve countries have decided to adopt the DVB-T system for their digital broadcasts.

VI   HDTV, the future and related topics

16.0   High definition television (HDTV)

Digital terrestrial television broadcast (DTTB) is a technology that enables more efficient delivery of television services than the analogue system of today. The following are the main benefits of the technology:

- Better reception
- Better quality pictures and sound
- Provides a number of standard definition channels within the same bandwidth needed to transmit one analogue channel
- Can provide one high definition channel within the same bandwidth needed to transmit one analogue channel
- Information of other types can be simultaneously carried, ie various audio data for different languages, text data, still images etc.
- Less transmission power needed.

High definition television is one of the main reasons for moving towards a digital transmission medium. There simply is not enough bandwidth in the analogue domain to allow for HDTV transmission. It is clear that when analogue transmissions are finally switched off and digital is the norm, that high definition (HD) will soon take over from standard definition (SD). One of the main reasons that HDTV is not taking off today is due to the cost of the screen. Although technologies are constantly improving, at the time of writing this book, they are not yet available in the mass consumer price range. HD offers a far better image resolution and quality than SD. The image quality, depth of detail and sheer pleasure to watch are incomparable. The HD definition also makes the HDTV a suitable vehicle for the delivery of other multi media services such as Internet, games, and video telephony.

The functional blocks necessary to realise HDTV are shown in *figure 16.1* and will be described now.

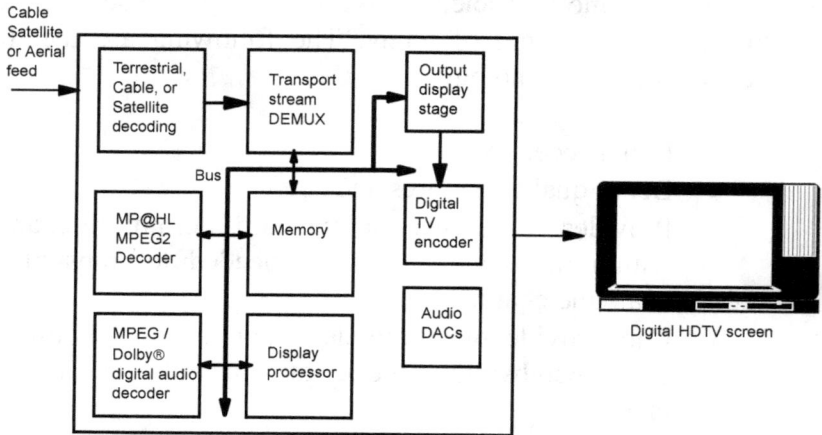

Cable
Satellite
or Aerial
feed

Terrestrial, Cable, or Satellite decoding	Transport stream DEMUX	Output display stage
MP@HL MPEG2 Decoder	Memory	Digital TV encoder
MPEG / Dolby® digital audio decoder	Display processor	Audio DACs

Bus

Digital HDTV screen

Figure 16.1

16.1   Terrestrial, cable or satellite decoding

As can be seen from *figure 16.1*, the input stream is from either a standard TV aerial, a cable or a satellite feed. Digital programming is available from all these sources. So after tuning to the appropriate frequency, the front end (FE) decoding is then performed. So this will be either QAM, QPSK, CODFM or 8-VSB.

16.2   Transport stream DEMUX

The next function to be performed is the demultiplexing. This simply strips out the programming material (audio/video) and associated data from the transport stream. The audio and video decompression / decoding is then performed. The critical aspects of this processing are performed by dedicated hardware blocks inside the DEMUX block. However, control of these, as well as many other tasks are generally performed by a high performance 32 bit RISC or VL-RISC micro.

The other tasks performed are:

- Construction of graphics for the OSD
- Control of the Remote control unit
- Control of the DEMUX processing
- Control of the MPEG and Dolby® decoding
- Control of the FE

16.3 MPEG / Dolby® decoder

For HDTV the MPEG 2 decoding will be main profile at high level (MP@HL). See section 3.1 of this book for an explanation of this. There are many standards available for the compression and subsequent decoding of audio data. A very popular one for many consumer applications including HDTV is Dolby® AC3. More detail on this is given in section 17.0 of this book.

16.4 Memory

Memory is needed for a number of purposes. The type used is usually synchronous dynamic random access memory (SDRAM). This is used to store the video and audio data while the MPEG processing is performed. Typically the size needed to handle HDTV pictures at resolutions 1080 lines (interlaced) by 1920 pixels is 128 Mbits, and 720 lines (progressive) by 1280 pixels is 64 Mbits. It is also used to store the micro code for the high end processor, who's functions are described under the DEMUX block. FLASH memory is also needed to store the micro code program. This will typically be around 16Mbits.

16.5 Progressive Scan

Progressive scan TV's and DVD's with progressive scan output and now becoming popular. A TV can write data onto its screen either in the conventional interlaced way or progressively. Interlaced means that the scanning lines will be drawn initially only to cover half the screen, ie missing out every other line, as shown in *figure 16.2:*

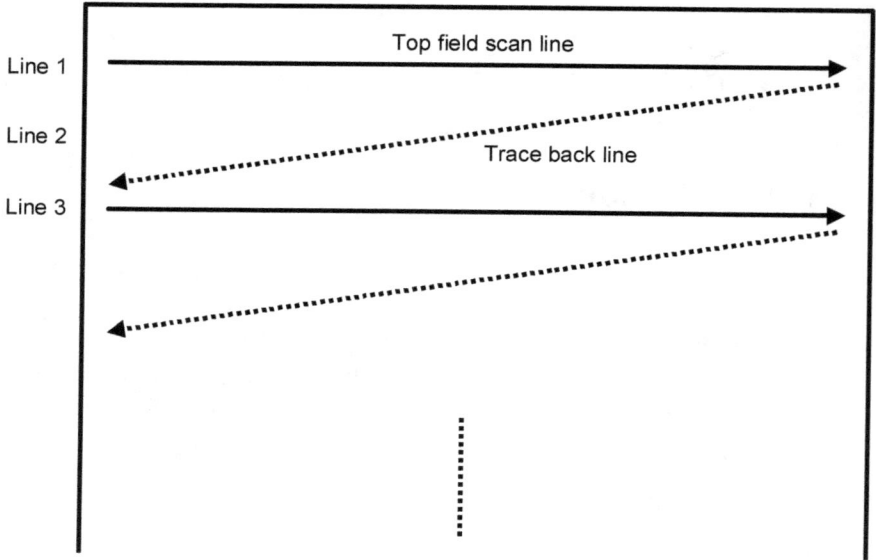

Figure 16.2

Then, the scan line will fly back to the top again and draw the lines that were missed out the first time, as shown in *figure 16.3.*

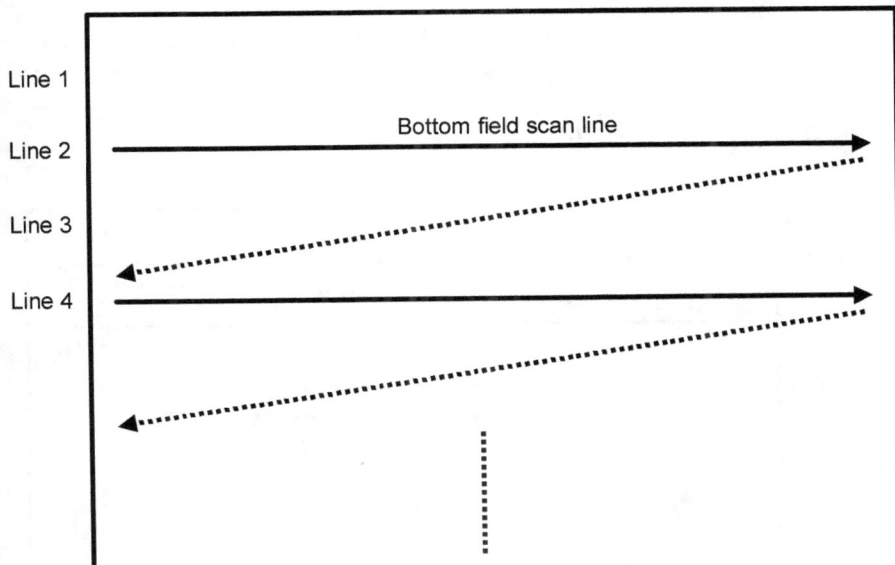

Line 1

Line 2   Bottom field scan line

Line 3

Line 4

Figure 16.3

Due to persistence of vision the fact that the TV does this is not always seen. Although with larger screens a certain amount of flicker is actually visible. It is for this reason that progressive scan is now also available, and operates as shown below, in *figure 16.4*:

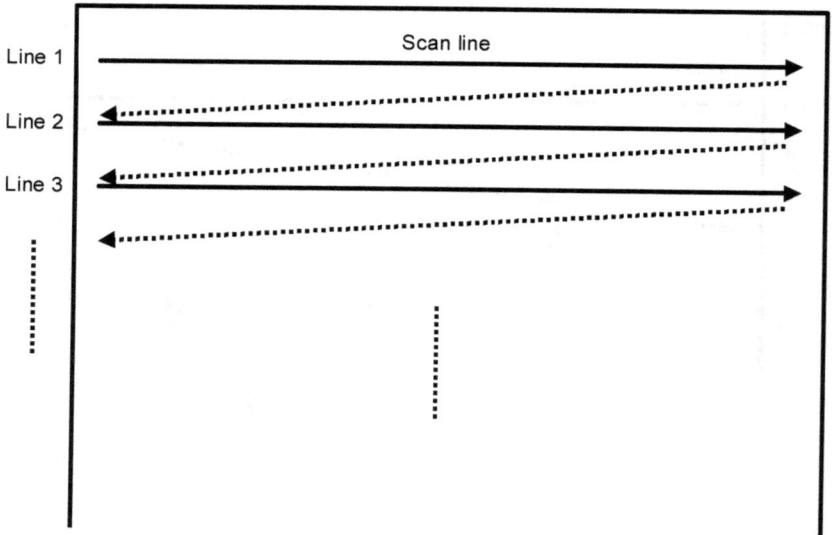

Figure 16.4

Incidentally, for SD TV there are 625 lines in total for European TV's and 512 for US TV's. (Note that not all these lines are visible). TV's were first developed with progressive scanning. However, due to the technology available at the time the refresh rate was so slow that the flicker could easily be seen as one screen picture was drawn and then the other. The interlacing method was then introduced to double the frequency rate; 50Hz in Europe and 60Hz in the US, without having to increase the transmission bandwidth (Bye the way, it's no coincidence that these frequencies matched the mains power distribution frequencies in the respective countries). Interestingly now it is the progressive scan system that gives much better and flicker free viewing.

16.6   Display processor

Once the decoding has been performed the data is passed to the display processor. Here various filters perform transformations to allow for the various possibilities of input and output formats. For example the video input may be SD resolution, for display on a HDTV screen. The input may be HD for display on a SD screen. The viewer may wish to view in HD and record to a standard VCR in SD etc.

16.7   Output display stage

Once the Video is in the correct format, then the multi media features need to be added, ie the OSD menu system, picture in picture etc. The OSD will need to be mixed in with the video, it may be necessary to have the video still be seen behind the graphics and so different levels of transparency need to be defined for the OSD. All these types of function are performed here. To produce good quality OSD graphics, a separate graphics processor may also be needed and also 24 bits per pixel quality for true colour rendering.

16.8   Digital TV encoder (DENC)

Once the final output picture has been constructed it simply needs to be sent to the TV in the correct format. Generally there could be any of 6 different output to the TV, these are:

- S-VHS
- CVBS (Composite video)
- YUV
- RGB (3 DAC's)

Hence 6 video DAC's (each generally 10 bits wide) are needed inside the DENC to convert the digital stream into analogue for output to the TV screen. The analogue formatting is also done, to achieve PAL, NTSC or SECAM signal outputs.

16.9   Audio DAC's

There will generally be 6 audio DAC's each 24 bits wide. This is to convert the Dolby® 5.1 channel digital audio data to analogue to drive the front left, front right, back left, back right, centre and sub woofer speakers. Other audio modes that may be supported are Dolby® Prologic, DTS, MPEG audio, and MP3. The DAC's will generally be operated at 32KHz, 44.1KHz, 48KHz, 96KHz, 176.4KHz or 192KHz.

16.10   Additional blocks

Depending on the particular implementation of the HDTV a number of other functions may also be implemented. These are the inclusion of a modem for PPV and VOD ordering and billing, also for internet access. A conditional access card slot may also be included for the viewing of certain programming that requires a monthly subscription. This may take the form of a dedicated slot and card, or of a common interface, a standard that has been put forward for terrestrial viewing in Europe. If internet viewing is required then a more powerful processor will also be needed to perform all the low level protocol, as well as the higher level HTML and Java processing functions.

17.0　　　　　Audio 'standards'

In the whole area of digital services the MPEG-2 standard has been adopted for the video, however no such agreement exists for the audio coding technology. The most common ones used today are MPEG audio, Dolby® digital, PCM, SRS labs , or mixtures of various of these.

Aside from the audio compression system which is part of the MPEG 2 compression standard, there are others which will be described here, that are being used for STB applications as well as home music systems, DVD players, televisions, games etc.

17.1　　　　　Dolby® digital

Dolby® are a US organisation that have come up with a 5.1 channel digital surround sound system. This standard has been adopted as part of the ATSC standard for audio (also known as AC3). It is also being used world wide in many consumer applications such as audio amplifiers, DVD players, and satellite broadcast set top boxes. Generally a one off fee needs to be paid to Dolby® by the manufacturer of the consumer equipment, as well as a royalty per unit sold. The system supersedes the Dolby® Prologic 4 speaker system which has limited bandwidth, and not such good separation between the channels.

17.1.1　　　　　Dolby® digital system operation

The system can be operated in various different modes depending on how many speakers the end user has. However to get the best performance 6 speakers are needed, as shown in the diagram below:

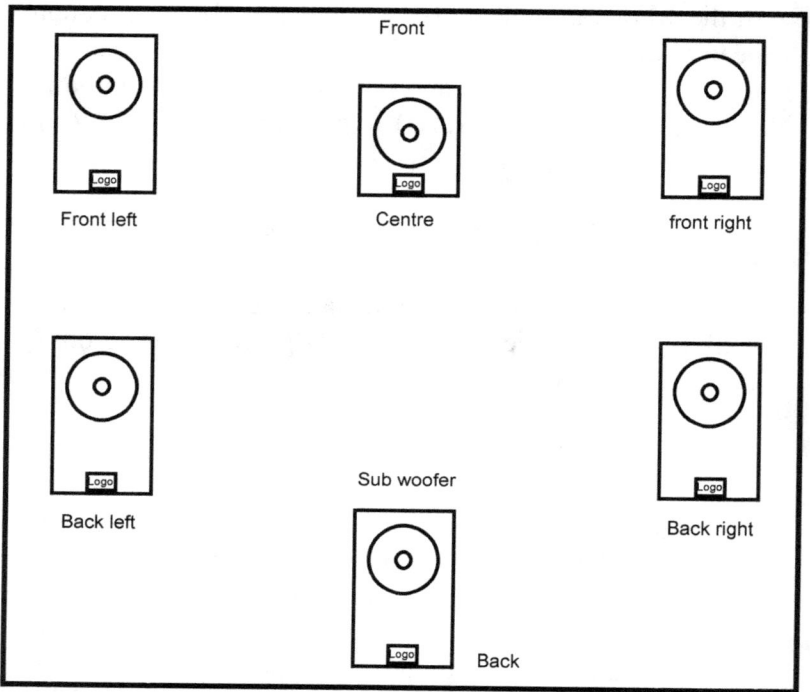

Figure 17.1

The system is also known as a Dolby® 5.1. This is because of the three speakers at the front, two at the back and an optional low frequency enhancement speaker. This speaker is only a fractional bandwidth speaker to carry the low frequency, or base signals. Hence it is referred to as 0.1 of a speaker.

The system works by first of all compressing the analogue audio signal. This is achieved by splitting the signal up into frequency bands. These bands are not generally of equal size, larger bands are possible for the lower frequencies. The compression is achieved based on perception of the human ear. For example if a particular signal amplitude is very close to a high one then the ear cannot hear the low one and so this low amplitude signal is not encoded.

The signal that is sent out to the end receiving equipment is composed of what is termed the *essence* information. This contains sound information and also some data. The second stream is termed the *metadata* This contains information regarding the essence data, as well as data related to the content of the audio data. The metadata will for example carry information regarding what is in the audio stream. It could be 5.1, stereo, or mono. A user may only have 2 speakers of 4. The receiver equipment will therefore decide how best to perform the down mixing. Further more this switching from 5.1 to stereo, to mono as required will be achieved with no level shifting. In fact the various options at the receiver that will require different types of down mixing are:

(i)   Mono
(ii)   Stereo
(iii) Prologic surround
(iv)   AC3 (5.1)
(v)   2 speaker virtual surround
(vi)   Headphone

Other information carried in the meta data allows for the control of the dynamic range. There is no standard at the programming level regarding at what signal level program audio should be set to. Hence different programs will have different levels. The speech level will also change with respect to other sound content within the same program. With the Dolby® system these problems can be overcome to give a uniformity of loudness acceptable to the listener. As a reference the level of the speech is taken. This is termed the *dialnorm* signal. Generally the dialogue level will be -16dB down from the rest of the audio signal for broadcast TV, and -25dB down for movies. Therefore when the viewer switches between channels there is no need to constantly change the volume settings on the TV.

17.2   SRS labs

SRS labs Inc. have done much research into the psycho acoustics of sound and the dynamics of the human ear, resulting in their so called head related transfer functions (HRTF's) the following standards were then released:

17.2.1   Sound retrieval system®

This technology takes stereo signals as input, breaks these down and produces a 3D sound effect as output. This output allows the listener to enjoy the stereo effect when walking around the room instead of only at a central point in the room.

17.2.2   TruSurround ™

This technology takes the mutli speaker signals from a technologies such as Dolby® digital (AC3), and produces a 3D sound effect from only two front facing speakers. It gives the listener the impression that there are additional, 'phantom' speakers around the room.

17.2.3   Circle surround ™

This technology takes as input a multi channel audio stream and encodes it into only two channels. These can then be transmitted and decoded at a receiver back into the original multiple channels.

18.0   The future, interactivity and interoperability

Generally speaking, the first wave of set top boxes and many around the world today are *closed systems*. That is to say that the software written to run on the set top boxes is non standard. A particular manufacturer's box will have some sort of application programming interface (API) that allows the low level hardware drivers to interface to a higher level application program. This high level application program, ie the electronic program guide (EPG), is then also non standard and is usually defined by the particular broadcaster. Also non standard is the conditional access system, which is also specified by the broadcaster. There will be a number of new products becoming available as these digital set top boxes become more widely distributed, such as: Games, internet, intranet, near video on demand, video conferencing, home shopping etc. The problem is that neither the content provider nor the broadcaster nor the set top box manufacturer will want to design these new products for each different box. The cost of doing this would price these new services too highly ensuring that no one would buy them. What is therefore needed is a common software platform much like Microsoft in the PC world. MHEG-5 as deployed on UK DVB-T boxes such an open standard for interactive STB applications. The term *interoperability* is the word generally used to describe the development of new services software (specifically interactive services) that can be targeted to run on all future set top boxes. To the viewer this means that these new interactive services will be able to run on which ever service he or she subscribes to, be it terrestrial, satellite or cable, also on which ever manufacturer's box they have. At the writing of this book, the middleware knows as MHP (Multi-media Home Platform) is set to be the world wide standard for interactive STB's be they DVB-T, DVB-C or DVB-S.

18.1   Advanced STB architectures

As silicon technology advances it becomes feasible to add
more and more features onto the basic STB central processor.
Figure 18.1 below shows a typical architecture of *a personal
video recorder* (PVR).

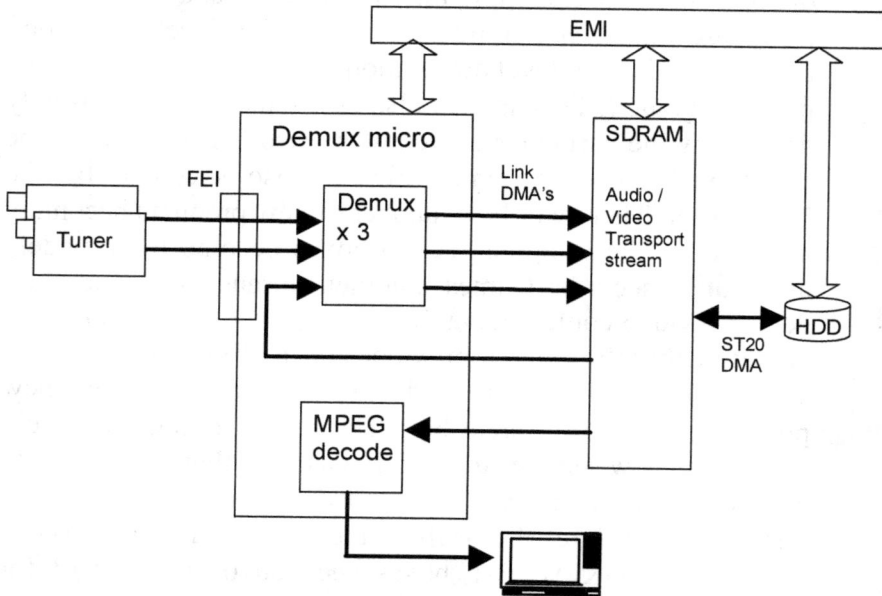

Figure 18.1

Figure 18.1 shows a solution based on one of
STMicroelectronics devices (STi5514). This device has three
demux engines and so such a system allows for viewers, for
example, to view one program with time shift capability whilst
at the same time being able to record a completely different
program from another input. (Note that in the
STMicroelectronics devices the EMI is the External Memory
Interface, see section 9.1.1 of this book for more information).
The demuxing is therefore performed on two transport stream
inputs from the tuners, and one from the *hard disc drive*
(HDD). By storing the PES data to disc, then watching the
movie by decoding this disc information it is possible to (1)

take a break, then carry on viewing the movie when ready, (2) skip the commercials by fast forwarding or skipping past them, or (3) going back and seeing a piece of the movie again even in slow motion. Clearly the HDD can be used also for standard movie recording. (Clearly the copyright issues need to be clarified). Typical storage times with different size discs are shown here:

HDD capacity	Recording time
10 GB	4 hours 30 minutes
20 GB	9 hour
30 GB	13 hour 30 minutes
60 GB	27 hours

Assumptions:
- Typical data rates are from 4 Mb/s to 6Mb/s

Table 18.1

Table 18.2 below shows the typical features available for a system incorporating a HDD.

Time shift features	Pause	Graphic display of elapsed time
	Fast backwards and forwards	Up to beginning and live. Graphic time display
	Zoom in and pan	In normal view, slow forward and pause
	Slow forwards	Slow backwards performance to be verified
	Next frame (in pause mode)	And previous frame
	Tag record as permanent	To archive with EPG data for later display. (A disc area is constantly dedicated to time shift)
PVR features	Record Now	
	Record by timer	Input service, data and start /stop time
	Record by EPG	Full schedule tables are displayed. User can sort by service, date, time and genre. The user selects an event, views the description and books the recording of the event in two touches.(One touch is patented). Software handles conflicts of bookings and viewing
	Select playback	From air as usual. From HDD by title, by record date by last viewed date, or all, or not watched records only
	Make play list	Select as for play back. Play play list sequentially of shuffle or one title only
	File management	Edit title, descriptors. Delete files
	Trick modes	Multiple fast / slow forwards / backwards / pause / skip (Similar to DVD trick modes, same as time-shift features)

Table 18.2

As HDD technology improves it will be possible to build video Jukeboxes similar to the audio Jukeboxes coming onto the market today. Multi input with dual MPEG decode systems will also give viewers a picture in picture (PIP) facility to allow a view of another channel when watching the primary one.

The advantage of HDD technology are not forgotten by the broadcasters either. They can charge veiwers for the capabilites to skip the advertisements and can also target specific advertisements to specific STB's storing advertising material to HDD.

18.1.1   Technology convergence

As a general trend, and for product differentiation reasons, more features are being 'bundled' in with the standard STB features of today. Broadband Internet is being made available for cable systems. Such features as DVD, CD-DA, MP3 and VCD functionality can also easily be integrated. The DVD data is simply stored in PES format, it has different error correction and decryption requirements, and also needs some navigation software. However STB's have already been developed today with no additional silicon being needed on the main processing board. *Figure 18.2* shows the basic architecture (Note that the demux processor must have a CSS descrambling block or capability to do this in software).

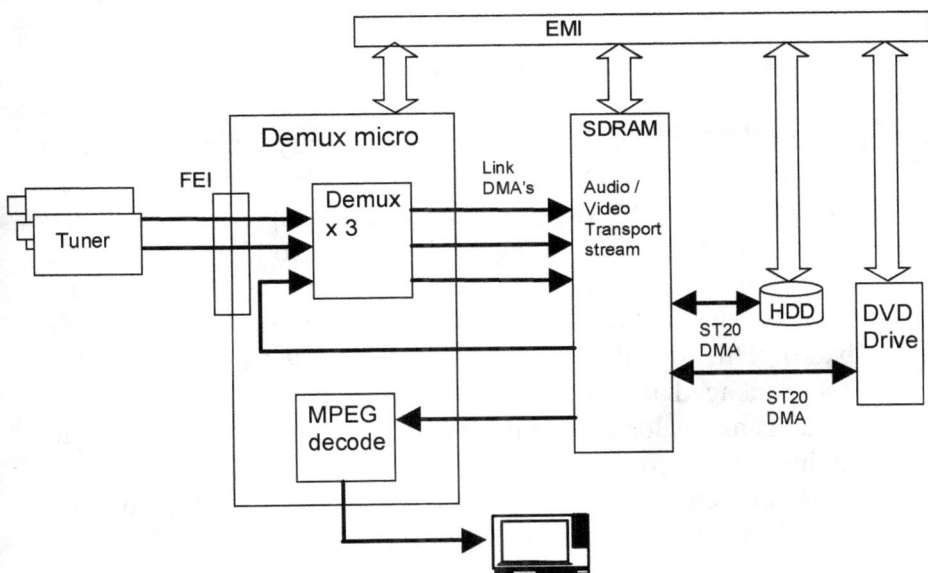

Figure 18.2

As has already been mentioned, the digital audio features will allow for impressive theatre sound effects. The STB can then be connected via an S/PDIF output to a home audio amplifier system. Although with the advances in digital audio amplification devices, the 6 channel amplification stages can also be integrated into the STB, as has already been done with some DVD players today.

The diagrams above show the tuner blocks outputting data in transport stream format. Devices are however available that have two COFDM demux engines inside the main demux processor thus given good cost reductions for such dual tuner systems.

18.1.2 CODEC systems

Figure 18.3 shows a block diagram of a fully integrated system with dual tuners, HDD for PVR functionality and DVD. It also included is an MPEG-2 coder. The example shown in the figure uses devices from Philips Semiconductors for the coding function. It allows an analogue input to be brought in. This would be the normal input to a TV (although at base band). This signal is then MPEG-2 encoded by the CODEC and packaged up into a transport stream. The audio is also input as an analogue signal (or can also, optionally, be brought in in S/PDIF format), where it is digitised and input to the coder device. Such a system is known as a CODEC solution since it has the ability to both code and decode the audio/video data.

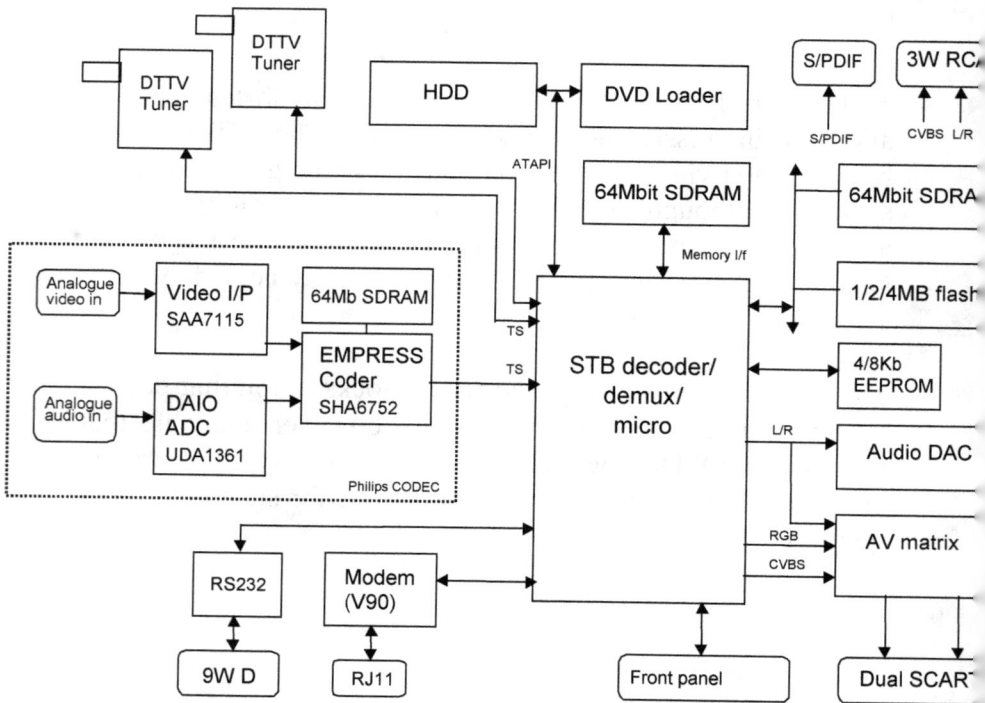

Figure 18.3

Such a system can clearly work for anyone irrespective of which CA system they are tied into.

The PVR functionality can then be enjoyed without the costs and / or restrictions imposed by the particular service provider, with just a penalty of very slightly degraded picture and sound quality. Such a fully featured system may also have a modem for interactive features.

The flash memory size will vary depending on the software requirements of the complete system. For example, for a UK system with MHEG-5, 4 Mbytes of flash will probably be needed. Note that a perfectly good system can be made without the MHEG-5 for the UK at the cost of not adhering to the recommendations of the DTG and BBC. A CI (Common interface) may be added to the above architecture should subscription services on terrestrial become popular again. Certainly some European countries will follow this model for some programming material. The RS232 port is generally used for update of the system software. However in such a system there are a number of ways that an upgrade to the software can be achieved:

(i)   From air (eg the Engineering channel in the UK)
(ii)   Via a CD ROM input to the DVD loader
(iii)   Via the V90 modem
(iv)   Via the RS232 port. This will more and more be replaced by USB interfaces
(vi)   Via the main demux processor programming / debug interface (such as JTAG).

Point (iv) is important since such systems will also be used for the input of digital camera images for viewing on the TV in slide show format.

Clearly the feasibility of such systems is very dependent of the software stacks available. At the writing of this book, these are available for exactly the examples shown in the last two subsections above.

19.0   Mobile DVB-T and DVB-H

Since the success of DVB-T and the increasing power levels used for transmissions, a new market area is fast emerging in Europe. At the writing of this book mobile and portable applications are already starting to appear in two main areas, (a) in car use and (b) connection via USB2.0 to the PC for personal TV's and PVR's. These applications however have very different technical requirements as will be looked at briefly in the following sections. The in car application of the standard DVB-T specification is the forerunner of the DVB-H applications which will for sure take over from DVB-T for such markets and spawn new ones in digital TV on the mobile phone and other hand held devices.

19.1   Diversity solutions

For the in car market the important criteria is for the system to tune to a frequency carrier while on the move. This therefore means the system will sometimes only have an echo to use instead of the main signal. It will also be moving through regions of different transmitters and so must also keep track of the best signal strength carriers to tune to. To aid in this task one of the main advances is the use of what is known as a diversity solution tuner. The diversity solution uses two antennas and two tuners to resolve the best signal to use at any given time. Generally each tuner signal must be demodulated individually, then the full COFDM decoding is performed on the best strength signal. At the writing of this book a leader in the field of demodulation of the diversity solution is a French company called "Dibcom".

19.2   DVB-H

The "H" in DVB-H is for Handheld. Hence the main focus is to reduce the power consumption of the receiving system, which in general will be a lightweight battery powered device. The standard will be used across the world including, it seems, the US. For the US the compression standards for the content are currently split between MPEG-4 and Microsoft's VC-1. Only MPEG-4 is defined as part of the current DVB-H standard. Also new is the fact that IP data casting has been adopted to allow for the streaming of lower quality video and other data to hand held systems. Transmission of IP packets over MPEG-2 is based on digital storage media command and control (DSMCC) sections that are typically not repeated. The transparent transmission of IP is accomplished by encapsulating the entire IP packet payload within the payload of a DSMCC section and by mapping the MAC address to the respective header and payload fields in the DSMCC section. The section format permits fragmenting datagrams into multiple sections.

The DVB-H asserts that the following features are supported:

- Low power for extended use
- Support for crossing from one transmitter region to another
- Scalable and flexible to allow for operation indoor, outdoor and while moving at various speeds
- Operation in high man made noise environments
- Operation in all regions of the world (ie flexible to various transmission bands and channel bandwidths)

How all these features are supported is by modifications to the normal DVB-T standard as explained here.

19.2.1   Power saving

Time-slicing is employed to reduce the power consumption of the hand held system. IF a large chunk of data can be sent, then nothing, the hand held device can be powered off for a period of time, until the next chunk of data is transmitted. The time-slicing is not performed on the PSI/SI data.

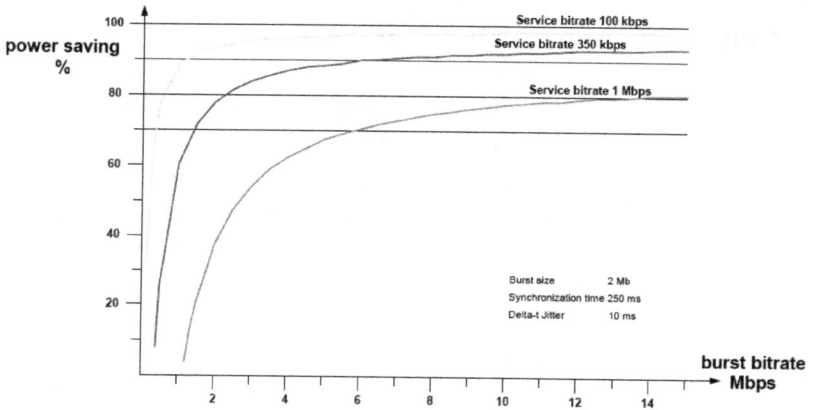

Figure 19.1

The figure above shows the DVB expectations of power saving for different bit rate applications. In reality these 90% power saving figures may be difficult to realise.

19.2.2   Transmitter crossover

During the power off phase it is possible for the receiver system to monitor a neighbouring cell frequency, thus making a smooth handover possible. The cell ID is transmitted in the TPS field. (See section 5.5 of this book)

19.2.3   Noise immunity and Doppler

The forward error correction (FEC) is improved to combat impulsive noise. Additional Reed Solomon (RS) coding is made to IP datagrams. These RS bytes are sent over special

FEC sections. In fact there is a new FEC block known as the MPE – FEC. It manages flexible error correction and time interleaving. More time interleaving is needed as the frequency of the Doppler effect reduces at the receiver.

DVB-H data payloads are IP datagrams (or other network layer datagrams encapsulated into MPE sections)

19.2.4   Flexibility

A new 4k mode with 3409 active carriers (in addition to the normal 2k and 8k modes) is incorporated to offer an additional trade off between SFN, cell size and mobile reception. As the DVB put it:

"The 8k mode can be used for both single transmitter operation and for small, medium and large SFNs. It provides a Doppler tolerance allowing *high speed* reception"

"The 4k mode can be used for both single transmitter operation and for small and medium SFNs. It provides a Doppler tolerance allowing *very high speed* reception"

"The 2k mode is suitable for single transmitter operation and for small SFNs with limited transmitter distances. It provides a Doppler tolerance allowing *extremely high speed* reception"

As with DVB-T, DVB-H can be used in 6, 7 and 8MHz channel bandwidth environments. However a 5MHz option is also available (for DVB-H for use in non broadcast environments). DVB-H can also co-exist with normal DVB-T transmissions within the same multiplex. Ie a broadcaster can choose to transmit two DVB-T services or one DVB-T service and one DVB-H service in the same multiplex.

VII   Standard Theory

20.0   Complex numbers

The most convenient notation for expressing waveforms in communications theory in general is in terms of complex numbers. This section is included as a revision of this area of mathematics.

20.1   Phasor representation

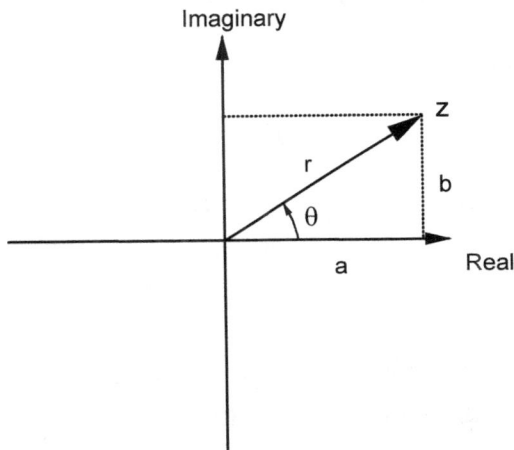

Figure 20.1

Figure 20.1 shows the representation of a complex number. Complex numbers were first devised to allow for the representation of the roots of negative numbers. ie:
$\sqrt{-1}$ .This is represented as j. In general a number will have both a real part and an imaginary part. This imaginary part is shown as being on another axis, as the diagram above indicates. Some texts use the letter 'i' for imaginary. In this book we will use 'j'. (This is so as not to confuse with I and Q notation when describing data relating to QAM modulated signals).

The number z is therefore complex since it has a real part and an imaginary part. ie $z = a + jb$.

Z can also be represented as being a vector with magnitude r.

Hence:

$$r = \sqrt{a^2 + b^2}$$

Also by using basic trigonometric identities:

$$Tan\,\theta = b/a$$

$$b = r\,sin\,\theta \qquad a = r\,cos\,\theta$$

and since z = a + jb:

$$z = r\,(cos\,\theta + j\,sin\,\theta)$$

This can more conveniently be represented as:

$$z = re^{j\theta}$$

(This can be shown to be true by comparing series expansions of e^{θ} with those of $cos\theta$ and $sin\theta$).

The importance of this identity in communications theory is that a waveform can be considered to be a vector, as shown above, with r representing its amplitude and θ its phase.

20.2   Waveform notation

Using the aforementioned notation, if the vector rotates with angular velocity ω (units in radians), then a continuous waveform (CW) like a sine wave can be defined as below, where ω is the frequency, t is time, Ø is the phase, and A is the amplitude.

$$CW(t) = Ae^{j(wt+\emptyset)}$$

The phase has been split here into two parts. A part that varies with time (ω), and a part that is 'pure' phase (Ø). It is useful to do this since we can now use this notation to compare signal phases, even with signals of the same frequency. Note that ω is in radians. To use the frequency (f) in the normal way, ie simply in Hz, then ω is replaced by $2\pi f$ to give:

$$CW(t) = Ae^{j(2.\pi ft+\emptyset)}$$

21.0　　　　Modulation techniques

Generally speaking modulation is the process of changing an original signal by a second one. The purpose of modulation techniques is two fold: Firstly they allow data to be effectively communicated. For example FM or AM radio. To allow reception of a signal with a reasonable size aerial the data must be transmitted within a certain frequency band. Hence a data carrying waveform is modified by the modulating carrier frequency. Secondly it is possible to increase the efficiency of the transmission by various modulation techniques. It is this feature that is important in the DVB-T system. The modulation techniques used in DVB-T are QPSK, 16-QAM and 64-QAM. These techniques are used in many analogue communications systems but also for COFDM, where digital modulation methods only are used.

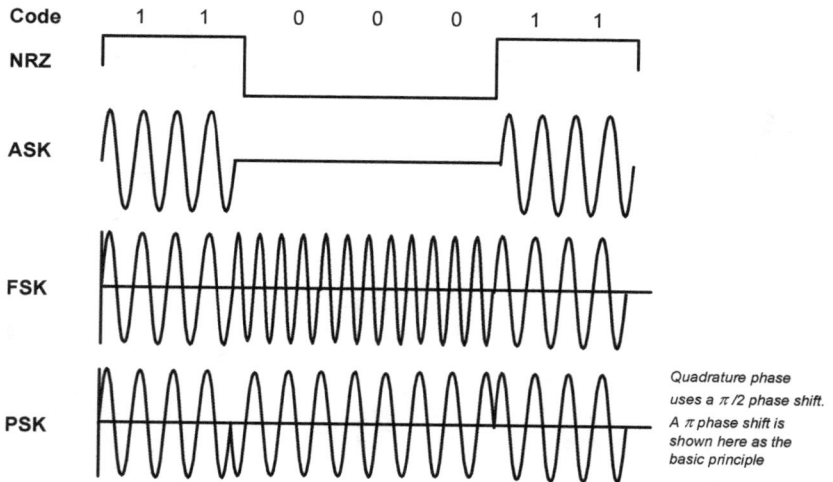

Figure 21.1

Figure 21.1 shows some basic definitions. Data to be transmitted; be it a transport stream or any other digital data stream is shown as the top. The first waveform is a representation of the bit pattern. It is known as non return to zero (NRZ) code since after representing the bit '1' as a high voltage level, the voltage does not return to the zero voltage level. Instead it only changes again in response to the next bit to be represented. If the next bit is another '1', the voltage will remain high, if the next bit is a '0', the voltage will go down to the low (zero or negative) voltage level. Now to allow this to be transmitted somewhere there are a number of ways that this 'high/low level' signal can be modified.

21.1   Amplitude shift keying (ASK)

The next waveform shows the amplitude shift key (ASK) signal. Here a second signal of fixed frequency is used. A simple addition of the two waveforms yields the ASK signal. This only has a burst of the second fixed frequency signal when the high level voltage is present, and no signal when the low level voltage is present. The NRZ code signal is said to modulate the fixed frequency signal.

21.2   Frequency shift keying (FSK)

An alternative method is shown in the next waveform down of *figure 21.1*. Here the second waveform used is of constant amplitude, but varies in frequency depending on the nature of the signal that is modulating it. A particular frequency is used for the high level and a different one for the low level.

21.3   Phase shift keying (PSK)

The next waveform down shows a method of modulation using the phase of the high frequency signal. The frequency is changed in phase depending on the high or low level of the first waveform. The figure shows the case where the high frequency waveform can adopt only one of two phases. A waveform's phase can be modified anywhere between 0 to 2π radians (0^0 to 360^0). The figure shows a waveform of a particular frequency representing a high level. It is then modified by π radians (180^0) to represent the low level.

However this method can be used to allow far more data to be transmitted if instead of one bit level (high or low) being represented by a particular phase, a group of bits can be represented by a particular phase instead. For example the following bit patterns can be associated with the following phases:

Bit pattern: 11   Phase: 0
Bit pattern: 10   Phase: $\pi/2$
Bit pattern: 00   Phase: π
Bit pattern: 01   Phase: $3\pi/2$

The bit pattern, 1100011 in *figure 21.1* would therefore be represented with the following phases: 0, $3\pi/2,\pi$, etc. In this example four different phases are used. This type of modulation is called *quadrature phase shift keying* (QPSK). In theory any number of phases can be used, however if the phases become too close together, and transmission noise is introduced, the receiver of the waveform will have a problem with decoding. In fact, to make decoding easier the amplitude of the high frequency waveform is also changed depending on the phase. This is now what is termed *amplitude shift keying* (ASK). It should really be called amplitude and phase shift keying, but is simply shortened. When four phases are used we have quadrature phase shift keying QPSK, however if the amplitude is also modified depending on the bit pattern we have a class of modulation termed *quadrature amplitude modulation* (QAM). The bit patterns shown here are correctly termed as *symbols*. With the addition of more phases and variations in amplitude, different numbers of symbols can be represented and hence transmitted. The terminology now used is n-QAM. Note also that in general for QAM the phases used are defined relative to some arbitrary waveform as +/- $\pi/4$, and +/- $3\pi/4$. This represents the same phase changes as shown above simply shifted by $\pi/4$, (ie $\pi/4$, $3\pi/4$, $5\pi/4$, $7\pi/8$). The symbols and there representations can now be more clearly represented on a phasor diagram. *Figure 21.2* shows a typical phasor representation of a QPSK modulated signal:

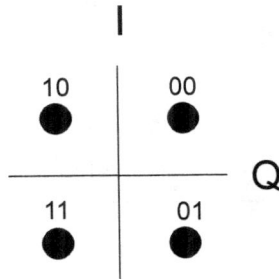

Figure 21.2

In this example the bit pattern 00 is represented with a phase of $\pi/4$, 01 by $3\pi/4$, 11 by $-\pi/4$, and 10 by $-3\pi/4$. The I and Q here represent; in phase (I), and quadrature out of phase (Q). This should not be confused with the complex number representation. *Figure 21.3* gives an example of a QAM representation. This phasor representation is actually known as a *constellation*. The figure in fact shows a possible *16-QAM* (or more accurately 16-PSK) constellation:

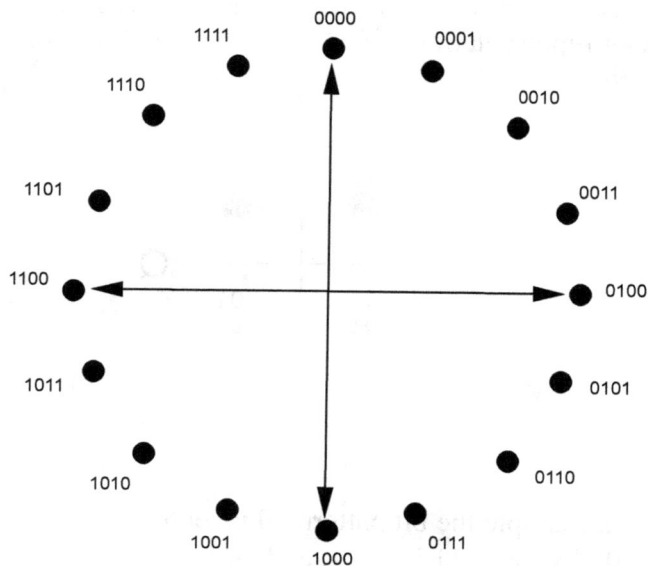

Figure 21.3

Although this is a constellation with 16 points and hence 16-QAM. As has previously been said, the amplitudes are modified to make reception easier. The true 16-QAM constellation is therefore shown in *figure 21.4*:

12 phases / 3 amplitudes in total

1 amplitude appears on 8 phases

2 amplitudes appear on 4 phases

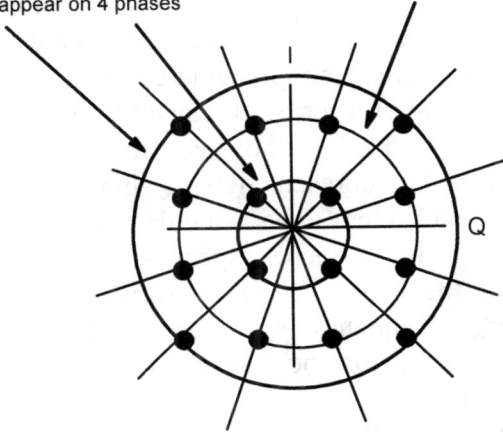

Figure 21.4

Here 16 symbols are represented by both phase and amplitude variations. Hence each symbol constitutes 4 bits. Note that there are only 12 phases in total. The '16' in 16-QAM is not related to phases alone, rather it is the number of symbols that are represented in total. Section 6.7 of this book shows the constellation diagrams for QPSK, 16-QAM and 64-QAM. These are the constellations used for DVB-T.

22.0   Fourier series

The Fourier series is important because it offers us a way of mathematically describing waveforms. Once this is achieved it becomes far easier to develop ideas and theories. This section will show you how a simple square wave function (infinitely long) is made up of a series of sine and / or cosine terms. This will then be extended to show you that a signal truncated in time can also be represented with sine and cosine terms of different frequencies. The calculation of this frequency representation, or frequency spectrum is in fact the *Fourier integral*, or the *inverse Fourier transform*. To then convert back from a frequency spectrum representation to a time varying representation the *Fourier transform* is used. Since a processor of some kind will be used to perform all the calculations in a real COFDM system, a method of calculating these Fourier transforms is needed, hence the *discrete Fourier transform* (DFT) is then looked at followed by the *fast Fourier transform* (FFT). Since these Fourier transforms are the central ideas of COFDM, this section will be fairly detailed.

22.1   Representation of functions by sine and cosine terms

Any periodic function can be represented mathematically by the following series:

$$f(x) = \frac{1}{2}a_0 + a_1 \cos x + a_2 \cos 2x \dots + a_n \cos nx$$

$$+ b_1 \sin x + b_2 \sin 2x \dots + b_n \sin nx$$

Equation (1)

ie a constant $a_0/2$ plus sine and cosine terms of different amplitudes, with frequencies that increase in discrete steps. This must meet the condition of convergence. ie it must not add up to an infinite value. This is the case for what are termed 'well behaved' functions. (Most signals in electronics can be considered to be 'well behaved').

We now need to find expressions for a_n and b_n. To do this we simply multiply both sides of equation (1) by cos mx, and integrate with respect to x over the period 0 to 2π. If this is done, we get:

$$\int_0^{2\pi} f(x) \cos mx \, dx = \frac{a_0}{2} \int_0^{2\pi} \cos mx \, dx$$

$$+ a_n \int_0^{2\pi} \cos mx \cos nx \, dx$$

$$+ b_n \int_0^{2\pi} \cos mx \sin nx \, dx$$

Now taking each one of these products in turn:

Looking at the a_0 term, clearly when integrated the cosine term leads to a sine term only: The sine of 0 and of 2π being 0. This term clearly vanishes.

Looking at the a_n term, it can easily be shown that:

$$a_n \int_0^{2\pi} \cos mx \cos nx \, dx = \left\{ \begin{array}{ll} 0 & if \quad m \neq n \\ a_n \pi & if \quad m = n \end{array} \right\}$$

This just leaves the b_n term. Again it can be shown that this is zero for all m and n. ie:

$$b_n \int_0^{2\pi} \sin mx \cos nx \, dx = 0 \quad for \; all \;\; m \;\; and \;\; n$$

This therefore leaves when m = n (we revert back to using n as the variable):

$$\int_0^{2\pi} f(x) \cos nx \, dx = a_n \pi$$

Or:

$$a_n = \frac{1}{\pi} \int_0^{2\pi} f(x) \cos nx \, dx$$

Equation (2)

Similarly, by multiplying both sides of equation (1) by sin mx and integrating from 0 to 2π we get:

$$b_n = \frac{1}{\pi} \int_0^{2\pi} f(x) \sin nx \, dx$$

Equation (3)

Now to find an expression for a_0. When n = 0:

$$a_0 = \frac{1}{\pi} \int_0^{2\pi} f(x) \cos 0 \, dx \quad = \frac{1}{\pi} \int_0^{2\pi} f(x) \, dx$$

So:

$$\frac{1}{2}a_0 = \frac{1}{2\pi} \int_0^{2\pi} f(x) \, dx$$

It can now be seen that the constant ($a_0/2$) at the beginning of equations (1), is just the average, or DC level (between 0 and 2π) on which the sine and cosine terms are superimposed. This will of course be zero if the waveform is symmetrical about the x axis.

So lets consider the mathematical representation of the following square wave using what we've just learned about the Fourier representation.

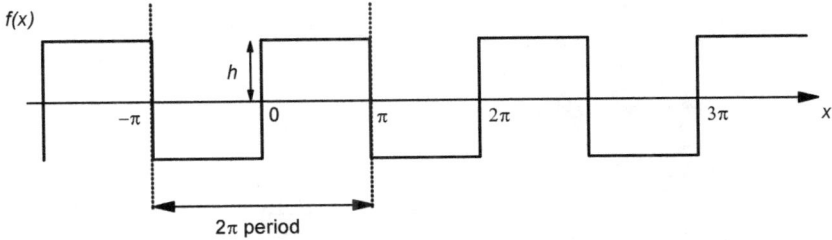

Figure 22.1

So the fist thing to do is to look at the coefficients, a_n and b_n, and calculate what these are.

Note that:

$$f(x) = h \qquad \text{between 0 and } \pi$$
$$f(x) = -h \qquad \text{between -}\pi \text{ and } 0$$

This is the mathematical definition of the square wave, f(x), shown above.

So b_n first:

$$b_n = \tfrac{h}{\pi} \int_0^\pi \sin nx \, dx - \tfrac{h}{\pi} \int_{-\pi}^0 \sin nx \, dx$$

$$= \frac{h}{n\pi}\{[\cos nx]_\pi^0 - [\cos nx]_0^\pi\}$$

$$= \frac{h}{n\pi}\{(1 - \cos n\pi) - (\cos n\pi - 1)\}$$

$$= \frac{2h}{n\pi}(1 - \cos n\pi)$$

So if n is odd $b_n = 4h / n\pi$
or if n is even $b_n = 0$

Therefore, as shown in *figure 22.2* our square wave can be represented as:

$$f(x) = \frac{4h}{\pi}(\sin x + \tfrac{1}{3}\sin 3x + \tfrac{1}{5}\sin 5x + \ldots\ldots)$$

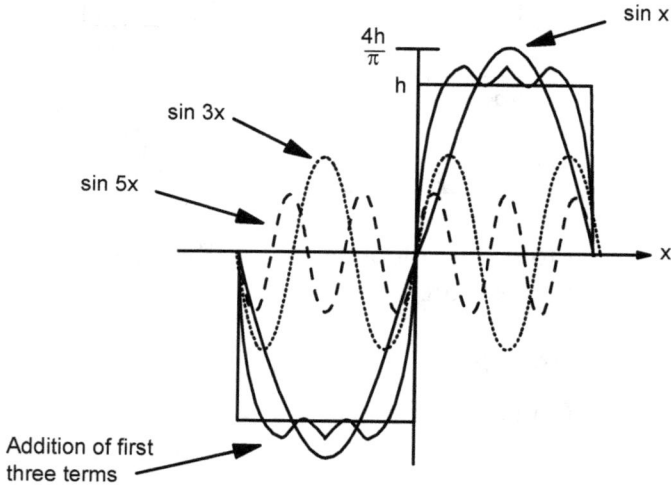

$$f(x) = \frac{4h}{\pi}(\sin x + \tfrac{1}{3}\sin 3x + \tfrac{1}{5}\sin 5x + \tfrac{1}{7}\sin 7x + \ldots.)$$

Figure 22.2

The sin*x* term is known as the fundamental frequency, the subsequent terms are the first and second harmonics.

22.2   Derivation of the continuous Fourier transform

We are now interested in knowing how we can treat time varying signals and only small chunks of these rather than a particular one being repeated for an infinite time. As is explained in section 7.1, a COFDM transmission system breaks up the time varying signal into chunks known as *symbols* for transmission.

Lets go back to equation (1) This can be re-written as:

$$f(x) = \tfrac{1}{2}a_0 + \sum_{n=1}^{\infty}(a_n \cos nx + b_n \sin nx)$$

$$= \tfrac{1}{2}a_0 + \sum_{n=1}^{\infty} c_n \cos(nx + \theta_n)$$

Equation (4)

Where

$$c_n^2 = a_n^2 + b_n^2$$

and

$$\tan \theta_n = b_n / a_n$$

Or alternatively:

$$f(x) = \sum_{n=-\infty}^{\infty} d_n e^{jnx}$$

Equation (5)

Where

$$2d_n = a_n - jb_n \quad (n \geq 0)$$

and

$$2d_n = a_n + jb_n \quad (n < 0)$$

This is the complex number representation.

Note that we have purposely arranged for n to range from $-\infty$ to $+\infty$, ie symmetrically about the imaginary axes.

To make this more useful in describing real signals found in electronics, we will replace the variable x with ωt, to give:

$$f(t) = \sum_{n=-\infty}^{\infty} d_n e^{jnw_0 t}$$

Equation (6)

Note that ω_0 is the fundamental frequency and must be in radians.

Substituting for a_n and b_n in the identity for d_n , ie equations (2) and (3) into equation (5), gives:

$$d_n = \left\{ \tfrac{1}{\pi} \int_{-T/2}^{T/2} f(t) \cos nw_0 t \; dt - j\tfrac{1}{\pi} \int_{-T/2}^{T/2} f(t) \sin nw_0 t \; dt \right\}$$

$$d_n = \frac{1}{T} \int_{t=-T/2}^{T/2} f(t)e^{-jn\omega t}dt \qquad \text{Equation (7)}$$

for all integer values of n.
Note: T is the period and has replaced 2π.

Converting ω from radians means $\omega = 2\pi f_1$, where f_1 is the fundamental frequency. ie the first frequency in the sine and cosine series. (The harmonics being integer multiples of this). So simply substituting equation (7) into equation (6) gives:

$$f(t) = \sum_{n=-\infty}^{\infty} \left[\frac{1}{T} \int_{-T/2}^{T/2} f(t)e^{-j2\pi n f_1 t}dt \right] e^{j2\pi n f_1 t}$$

The next steps in deriving the Fourier transform are as follows:

1)   Let T approach infinity. (This isolates a single pulse, ie we are saying that in an infinitely long period there will only be one pulse).

2) $f_1 = 1/T$, as $T \to \infty$ $1/T \to$ becomes infinitesimally small and can be written as df

3) nf_1 can be written simply as f, when n becomes large and $1/T \to 0$

4)   The sum over n can now be replaced by an integral since a unit change in n produces an infinitesimal change in
$n/T = nf_1$.

This gives:

$$f(t) = \int_{f=-\infty}^{\infty} \left[\int_{t=-\infty}^{\infty} f(t)e^{-j2\pi f t}dt \right] e^{j2\pi f t} df$$

224

or

$$f(t) = \int_{f=-\infty}^{\infty} F(f)e^{j2\pi ft}\, df \qquad \text{Equation (8)}$$

This is the *Inverse Continuous Fourier Transform* (**ICFT**), or *Fourier Integral*.

where

$$F(f) = \int_{t=-\infty}^{\infty} f(t)e^{-j2\pi ft}\, dt \qquad \text{Equation (9)}$$

This is the *Continuous Fourier Transform*. (**CFT**).

With the amplitude of the coefficients given by:

$$d_n = \tfrac{1}{T} \int_{t=-T/2}^{T/2} f(t)e^{-jn\omega t}\, dt$$

Note that the ICFT becomes:

$$f(t) = \tfrac{1}{2\pi} \int_{\omega=-\infty}^{\infty} F(\omega)e^{j\omega t}\, d\omega$$

if $2\pi f$ is replaced by ω.

The Fourier transform therefore represents a non-periodic wave group. Integration with respect to frequency produces a function in time. So given a function that is distributed in time, we can now express it as a spectrum of frequencies. Conversely, the inverse Fourier transform allows us to express a frequency spectrum, as a function distributed in time.

22.3   Example application of the IDCT

As an example of how the DCT is used, lets consider the following voltage square wave:

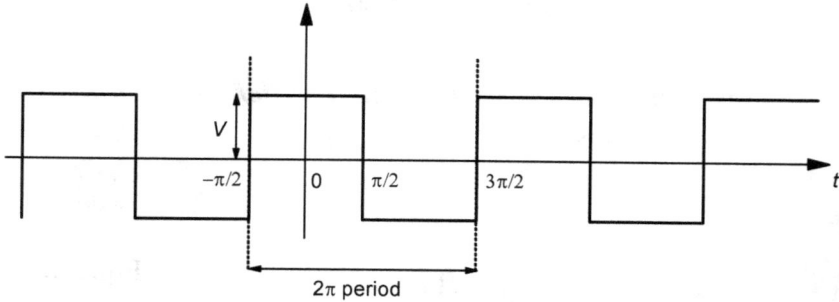

Figure 22.3

It has already been said that this signal can be described by a series of sine and/or cosine terms. Simply by looking at the waveform we can see that only the cosine terms will be needed to approximate it. But lets go through the calculation to verify this. The function $f(t)$ can be described as shown in equation (1) ie:

$$f(t) = \frac{1}{2}a_0 + a_1 \cos wt + a_2 \cos 2wt \ldots + a_n \cos nwt$$

$$+ b_1 \sin wt + b_2 \sin 2wt \ldots + b_n \sin nwt$$

(Where x has been replaced by ωt. Note $\omega = 2\pi/T$)

So we start by calculating what the coefficient should be. For this we use the previously derived equation (9):

$$d_n = \frac{1}{T} \int_{t=-T/2}^{T/2} f(t)e^{-jn\omega t}\, dt \qquad \text{Equation (9)}$$

For convenience the period is split into two. One half is taken from $-\pi/2$ to $\pi/2$, and the other from $\pi/2$ to $3\pi/2$.

So: $d_n = \dfrac{V}{T}\left[\displaystyle\int_{t=-\pi/2}^{\pi/2} e^{-jn\omega t}\,dt - \int_{t=\pi/2}^{3\pi/2} e^{-jn\omega t}\,dt\right]$

$= \dfrac{V}{T}\left\{\left[\dfrac{e^{-j\omega t}}{-jn\omega}\right]_{-\pi/2}^{\pi/2} - \left[\dfrac{e^{-j\omega t}}{-jn\omega}\right]_{-\pi/2}^{3\pi/2}\right\}$

$= \dfrac{V}{-j\omega T}\{e^{-jn\omega\pi/2} - e^{jn\omega\pi/2} - e^{-jn\omega3\pi/2} + e^{-jn\omega\pi/2}\}$

$= \dfrac{j}{2n\pi}[2e^{-jn\pi/2} - e^{jn\pi/2} - e^{-j3n\pi/2}]$

Note that $w = 2\pi / T = 1$ in this example,

And $\dfrac{1}{-j} = j$

We now expand the exponentials in terms of their sine and cosine term

$d_n = \dfrac{jV}{2n\pi}[\cos n\pi/2 - 3j\sin n\pi/2 - \cos 3n\pi/2 + j\sin 3n\pi/2]$

$= \dfrac{V}{2n\pi}[j(\cos n\pi/2 - \cos 3n\pi/2) + 3\sin n\pi/2 - \sin 3n\pi/2]$

Hence:

$2d_n = V[j(\cos n\pi/2 - \cos 3n\pi/2) + 3\sin n\pi/2 - \sin 3n\pi/2]$

            Equation (10)

Now d_n is in general a complex number as was seen previously, ie:

$$2d_n = a_n - jb_n \ \ (n \geq 0)$$

So looking at equation (10) the complex part of this becomes 0 for any value of n, ie for odd values of n, the cos terms become 0, and for even values they are equal and so cancel. This just leaves the real part, which, as expected describes the cosine coefficients. The sine terms in the equation are 0 for even values of n, and opposite values for odd values. Hence the coefficients are given by:

$$\pm 4V/n\pi \quad (+ \text{ or } - \text{ depending on } n)$$

This therefore gives the following frequency description of the signal $f(t)$.

$$f(t) = \frac{4V}{\pi}(\cos \omega t - \tfrac{1}{3} \cos 3\omega t + \tfrac{1}{5} \cos 5\omega t - \ldots\ldots)$$

Where $\omega = 2\pi/2$

22.4   The Discrete Fourier Transform (DFT)

To be able to realise the continuous Fourier transform and its inverse with digital circuits a modification has to be made. So lets take the ICFT first:

$$f(t) = \int_{f=-\infty}^{\infty} F(f)e^{j2\pi ft}\, df \qquad\qquad \text{Equation (8)}$$

As has just been seen, the ICFT is the representation of a *continuous* time function in terms of the integral of a *continuous* frequency function. These must be converted into *discrete* values to allow computational techniques to be used. The continuous function is sampled at a uniform sampling rate of $1/\Delta t$. N samples are used to approximate the continuous function. *Figure 22.4* shows the function $f(t)$ being approximated by $f(t_i)$, where t_i is an integer. The time variable t becomes $t_i\Delta t$. A similar treatment is performed for the DFT.

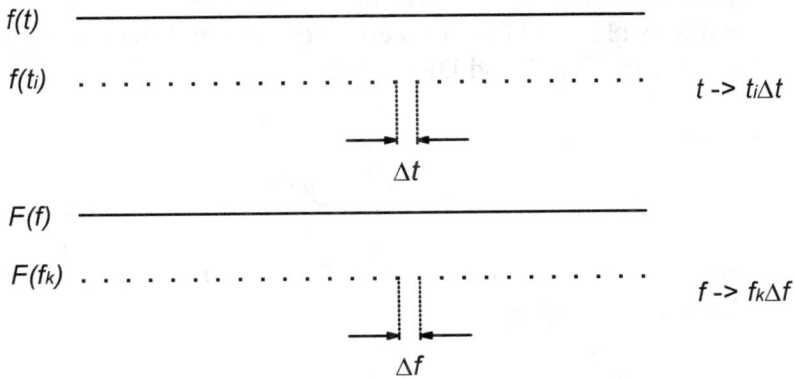

Figure 22.4

When an analogue signal is converted to digital, we need to know what sampling rate to use so that no information is lost. No loss of information occurs provided that when the digital signal is filtered, through a low pass filter (LPF), the resultant signal is identical to the original analogue signal. This is the case if the time between samples Δt is given by:

$$\Delta t \leq 1/2W \text{ seconds.}$$ Equation (11)

Where W is the bandwidth. W is the highest frequency component in the time varying signal.
The sampling rate is known as the *Nyquist sampling rate*, and is therefore:
$$1/\Delta t = 2W$$

However, with N frequency sampling periods, the highest frequency component is $N \Delta f$.
Therefore:
$$W = N\Delta f / 2$$

or $$\Delta f = 2W / N$$ Equation (12)

The integral is then replaced by the sum and the infinite limits replaced by finite ones covering the samples N. By substituting equations (12) and (11) into equations (8) and (9) respectively, the resulting IDFT and DFT's become:

$$f(t_i) = \frac{2W}{N} \sum_{f_k=1}^{N} F(f_k)e^{j2\pi f_k t_i/N}$$
Equation (13)

This is known as the **Inverse Discrete Fourier Transform (IDFT)**.

$$F(f_k) = \frac{I}{N} \sum_{t_i=1}^{N} f(t_i)e^{-j2\pi f_k t_i/N}$$
Equation (14)

This is known as the **Discrete Fourier Transform (DFT)**.

The coefficients can be calculated as follows:

Lets take equation (9)

$$d_n = \frac{1}{T} \int_{t=-T/2}^{T/2} f(t)e^{-jn\omega t}\, dt$$
Equation (9)

If we consider $f(t)$ as being made up on N discrete sampling points equally spaced out, at intervals of Δt then we get:

$$d_{f_k} = \frac{1}{T} \sum_{t_i=1}^{N} f(t_i\Delta t)e^{-j2\pi f_k t_i/N}.\Delta t$$

$$= \frac{1}{N} \sum_{t_i=1}^{N} f(t_i\Delta t)e^{-j2\pi f_k t_i/N}$$
Equation (15)

Note that $T = N\Delta t$.

Coefficient magnitudes can now be calculated. Note that the $f(t_i\Delta t)$ term is simply the result of a measurement of a point on the time varying waveform. Taking the waveform in section 22.3 as an example, this would simply be V between $-\pi/2$ and $\pi/2$, and $-V$ between $\pi/2$ and $3\pi/2$. For each coefficient the $t_i =$ 1 to N samples loop must be completed.

22.5　　　　The Fast Fourier Transform (FFT)

The FFT is a computational algorithm for calculating the DFT. Its aim is to calculate the Fourier transform much more quickly than simply calculating from the actual equations. It was developed by J. W. Cooley and J. W. Tukey in 1965. Since then the DFT, implemented with the FFT algorithm, has been widely used in many areas of engineering and science.

If the standard DFT is used, one would need to perform in the order of N^2 calculations (where N is the number of sample points over the period of interest). However the FFT allows this to be reduced to:

$$\frac{N}{2}\log_2 N \text{ calculations.}$$

With $N = 8192$, this means that if a typical calculation takes about 10 ns, the DFT would take 670 ms, but using the FFT algorithm, this would take 0.53 ms. An improvement by a factor of 1264.

Note that:　　$$\log_2 N = \frac{\log_{10} N}{\log_{10} 2}$$

22.6   Orthogonality

Two vectors are said to be orthogonal if the cosine of the angle between them is zero. This was explained in section 7.3 on COFDM principles.

Fourier methods allow us to represent signals in terms of periodic sine and cosine terms. These signals can be tested for orthogonality as with any vectors. Periodic waveforms are said to be orthogonal if they are laid out in the time axis and the average of the integral of the products of the pairs of values for all instants of time over their common period is taken, and is found to be zero. Putting this mathematically, two time dependant waveforms $S_m(t)$ and $S_n(t)$ are said to be orthogonal if:

$$\frac{1}{2T} \int_{-T}^{T} S_m(t).S_n(t) \; dt = 0 \qquad for \; m \neq n \; over \; 2T$$

23.0   Vestigial Side Band (VSB)

VSB is a form of amplitude modulation (AM) which was adopted as the US standard for digital television transmission in 1996. It was recommended by the FCC's advisory committee on advanced television services. The use of signals outside of the data carriers for synchronisation make this a rugged system, allowing for a synchronised well locked picture even if the data is corrupt. A consortium known as the *grand alliance* was formed in 1993 to lobby the FCC to adopt the VSB system for the delivery of digital standard resolution television (SDTV) as well as digital *high definition television* (HDTV). This consortium comprised of the following groups: AT&T, David Sarnoff Research Centre, General Instruments Corporation, Massachusetts Institute of Technology, Philips Electronics North America Corporation, Thomson Consumer Electronics, and Zenith Electronics Corporation). After running various trials they recommended the trellis coded 8-VSB system for digital terrestrial transmissions and 16-VSB for cable transmissions. This book describes the trellis coded 8-VSB digital terrestrial system in section 14.0.

To understand VSB we will first look at the fundamentals of amplitude modulation (AM).

23.1   Amplitude modulation (AM)

As is explained in section 20.0 on modulation techniques, the NRZ (Non Return to Zero) bit pattern modulates a carrier frequency for ASK (Amplitude Shift Keying).

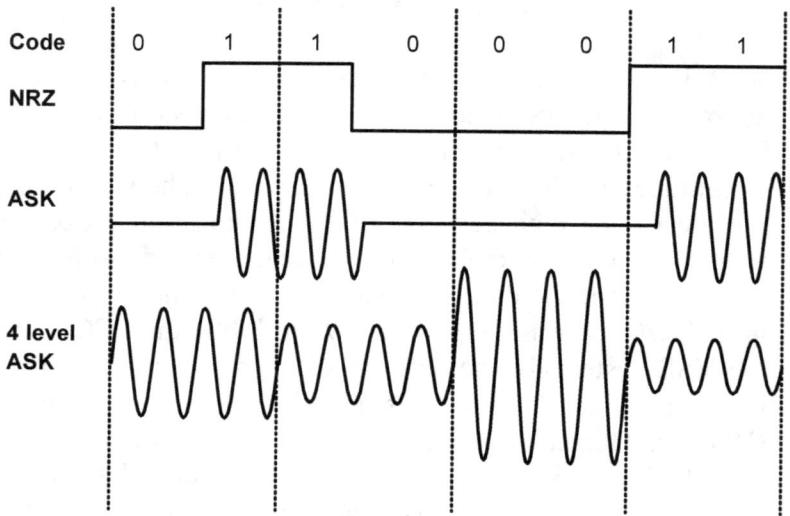

Figure 23.1

8-VSB uses a modulation technique very similar to ASK, but as shown in *figure 23.1*, it uses different amplitude levels for different bit patterns. In the example above four different levels are show for the bit patterns 01, 10, 00, and 11. The 8-VSB extends this idea to 8 levels, and so there are eight bit patterns (or more commonly known as *symbols*) each with three bits.

Lets now look at the mathematics of AM modulation. This will lead to an understanding of the frequency spectrum produced. We will then see the advantage of removing a component of this spectrum. This is the technique of VSB modulation.

Although *figure 23.1* shows the modulation of a carrier frequency by a square wave, we'll look at the general case of a single continuous wave modulating a carrier wave as shown in *figure 23.2*.

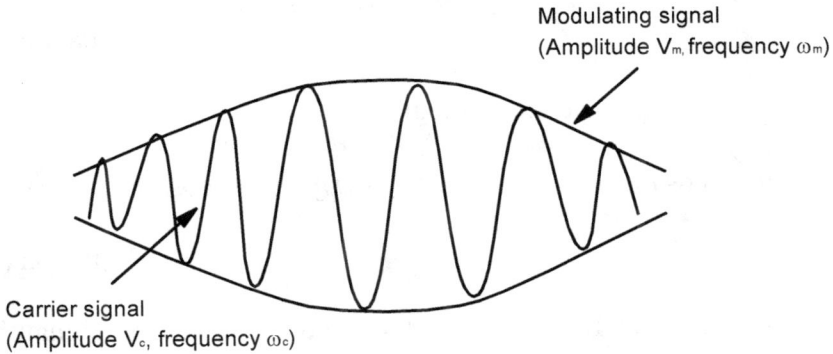

Modulating signal
(Amplitude V_m, frequency ω_m)

Carrier signal
(Amplitude V_c, frequency ω_c)

Figure 23.2

If a pulse is used to perform the modulation then the overall signal is given by the following expression:

$$V_C = (V_c + V_p) \cos(\omega_c t) \qquad \text{Equation (1)}$$

Where:

V_p is the modulating pulse voltage amplitude
V_c is the carrier voltage amplitude
ω_c is the carrier signal frequency in radians

Now by replacing the pulse amplitude with a modulating continuous wave, this gives:

$$V_C = [V_c + V_m \cos(\omega_m t)] \cos(\omega_c t) \qquad \text{Equation (2)}$$

Where:

V_m is the modulating pulse amplitude.
ω_m is the modulation signal frequency in radians

If we say that $V_m = M.V_c$ ie letting V_c be a multiple of V_m then we can rewrite the above equation as:

$$V_C = V_c \left[1 + M \cos(\omega_m t)\right] \cos \omega_c t$$

Equation (3)

Expanding this expression gives:

$$V_C = \cos \omega_c t + \tfrac{1}{2} M \cos(\omega_c t + \omega_m t)] + \tfrac{1}{2} M \cos (\omega_c t - \omega_m t)$$

Equation (4)

It can now be seen that the modulated carrier signal has components of its frequency spectrum at ω_c, at $\omega_c + \omega_m$, and at $\omega_c - \omega_m$. The two components on either side of the main carrier frequency are known as the *sidebands*. Plotting the frequency against power gives the following spectrum:

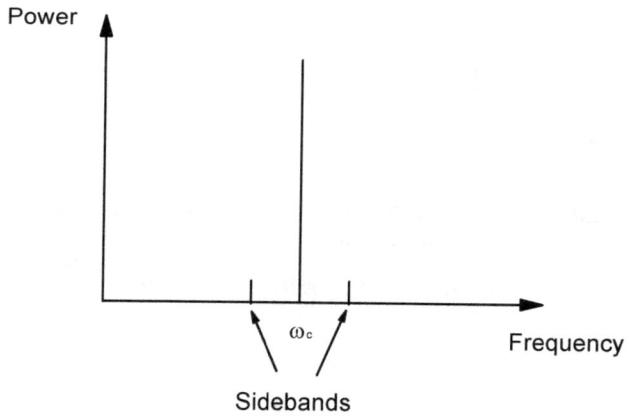

Figure 23.3

It can be seen that the majority of the power is in the central carrier frequency with only 1/16 th in each of the sidebands.

Generally the signal that is to be conveyed, for example a TV picture will be a complex waveform. This can be expressed simply as a function of time. Equation (3) can therefore be re-written as:

$$V_C = V_c[A + f(t)] \cos \omega_c t$$

Equation (5)

The general form of the spectrum is as shown here in *figure 23.4*:

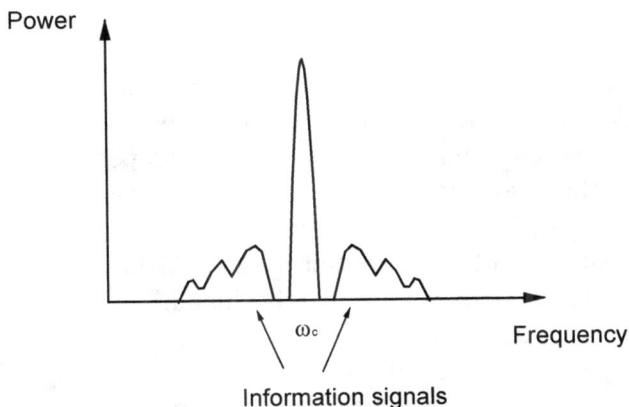

Figure 23.4

As can be seen, the low power sidebands are the information carrying signals.

23.2   Double side band suppressed carrier AM

Various methods can now be considered to reduce both the power and the bandwidth of a transmission system. For example, the carrier frequency can be suppressed with a large saving in transmitter power, with only the sidebands being transmitted. This is known as *double side-band suppressed carrier amplitude modulation* (DSB-SC-AM). Since the same information is transmitted in both sidebands then a reduction by half in the bandwidth needed can be made by simple transmitting only one sideband. This is known as *single side-band amplitude modulation* (SSB-AM). VSB is really a combination of DSB-SC-AM and SSB-AM. The carrier is suppressed and both sidebands are taken, however one sideband is reduced with only a vestige of it remaining. Hence the term *vestigial side-band modulation*.

24.0   Power and noise in communications systems

24.1   The decibel (dB)

Power is measured in terms of dB's (decibels). This term derives from the measurement of sound. Sound intensity can be measured in terms of number of watts per square metre. For example, two people standing next to each other would be talking at about 10^{-2} watts/m^2. If they decided to shout, the intensity would be increased by about 100 times. So changing by a few watts is completely inaudible. So a logarithmic measure was devised instead. When the sound intensity is changed by a factor of 10 it is said to have changed by 1 bel, a factor of 100 by 2 bels etc. A change of 1 bel is just perceptible by the human ear. In fact the human ear has a range of around 12 bels. The decibel is therefore 1/10 th of a bel. As well as being used for sound the dB is used throughout electronics and physics as the difference measurement of electrical power, as well as an absolute measure too.

24.2   The dB as a power measurement

Assume that two signals have power level given by P_1 $(=V^2/R)$, and P_2 respectively. The difference between these power levels is simply given by the expression:

$$\Delta = 10 . \text{Log}_{10} (P_1/P_2) \text{ dB}$$

P_2 in the equation above is commonly referred to as the reference power level. Ie power level P_1 is being compared with power level P_2. If a common reference power level is taken for convenience, then a signal can be measured to be a certain number of watts (W), or more commonly, miliwatts (mW) above (or below) this level. ie:

$$\Delta = 10 . \text{Log}_{10} P_1 \text{ dBm}$$

Hence the term dBm for dB above one milliwatt, or dBW for dB above one watt. Note that since $P_1 = V^2/R$, the definition can be re-written in terms of voltages as:

$$\Delta = 20. \, \text{Log}_{10} \, [(V^2_1/R)/(V^2_2/R_r)] \, \text{dB}$$
$$= 20. \, \text{Log}_{10} \, (V_1/V_2) - 10. \, \text{Log}_{10} \, (R_1/R_2)$$

Where V_1 and V_2 are developed across the resistances R_1 and R_2 respectively.

If $R_1 = R_2$ and $V_2 = 1$ then

$$\Delta = 20. \, \text{log}_{10} \, V_1 \, \text{dB}$$

24.3   Conversion between dBm to dBµV

In fact a common way to express power levels is in terms of voltage, usually in terms of dBµV. This is the number of µV a signal level is above (or below) another. Since measurements are sometimes given in dBm and sometimes in dBµV it is a good exercise to convert between them. As an example lets see what 0dBm corresponds to in terms of dbµV.

0dBm means the signal is 0dB above 1 mW, so it is *at* 1mW. Power (P) in watts (W) = V^2/R, hence power in miliwatts;

$$P_{mw} = V^2/R \times 10^{-3}.$$

Hence :

$$V = \sqrt{P_{mw}.10^{-3}.R} \quad \text{volts}$$

Or for V in µV

$$V = \sqrt{P_{mw}.10^{-3}.R} \, .10^6 \quad \mu V$$

So the power level in terms of number of dBs above 1dB expressed as µV is:

$$\Delta = 20.\log_{10}\left[\sqrt{P_{mw}.10^{-3}.R} .10^6 \right] \text{dB}\mu\text{V}$$

Note that an impedance term is in the expression. This is the impedance across which the voltage is developed. So the 0dB power, if dissipated across a 50 Ω impedance, can be expressed as:

107 dBµV (Remember that P_{mw} = 1 mw)

Or

109 dBµV in 75 Ω

Hence a typical signal which may be 60 dBµv in 75 Ω can also be expressed as:

(60 - 109) dBm = -49 dBm

(Ie the signal power level is 49 dB less than 1 milliwatt).

24.4   Bandwidth impact on power measurements

When performing measurements of power on signals with such equipment as spectrum analysers it is important to consider the bandwidth used to acquire the information.
Note the following to convert from a full power signal of S_{FB} MHz to one viewed at S_V MHz

The following number must be subtracted from the full bandwidth signal:

10 log (S_{FB}/S_V)

For example, if a 7.61 MHz bandwidth signal is at a full power level of -49 dBm, then if measured with an analyser set to a 300 kHz bandwidth, the analyser will indicate a power level of:

-49 - 10 log (7.61/0.3) = -63 dBm.

DVB Specifications Overview

[1]   Baseband Processing

DVB-MPEG [ETR 154]

This specification was drawn up to give a guideline on how to restrict certain syntax and parameters of the international MPEG-2 standards [ISO/IEC 13818 -1/2/3]. It gives preferred values for use with DVB services. It gives details on the parameters for HDTV and also the use of Dolby Digital audio within DVB.

DVB-TXT [EN 300 472]

Teletext has been a very popular service delivered by analogue transmissions for many years. The same service has therefore been retained as part of the DVB specifications. The delivery mechanism is described in this specification.

DVB-SUB [ETS 300 743]

When a program is broadcast in its original language many countries then transmit the local translation as subtitles. This specification allows the subtitles to be transmitted in digital, along with other graphical objects and logos.

DVB-SI [EN300 468]

Service information or SI contains information carried by the large number of services and programmes. For STB to be able to navigate and tune to the correct channels navigation aids are needed. These are transmitted as part of the DVB transport stream in the form of SI and described in this specification.
[EN 162] is also available to describe how the ST should or could be used. [ETR 162] SI codes are listed which indicate services by different broadcasters.

[2]   Transmission

DVB-S [EN 300 421]

This specification defines the delivery of services via satellite and was the first time the channel coding methodologies were defined. [TR 101 198] is also available describing DVB-S derivatives that use BPSK. (Normally QPSK modulation is used but under very severe interference conditions BPSK can be used).

DVB-C [EN 300 429]

This is the specification for transmission via cable (CATV). This also forms the basis for the [EN 300 473] specification (DVB-CS), which describes the use of Satellite Master Antenna TV installations (SMATV).

DVB-T [EN 300 744]

This specification is the main focus of this book for digital terrestrial transmissions. Guidelines are also given in [TR 101 190].

DVB-SFN [TS 101 091]

For the case of a single frequency network approach to DVB-T, the synchronisation of the transmitters is very important. The details of this are given in the specification entitled 'Mega-frame for SFN synchronisation'.

DVB-H [EN 302 304]

Digital transmission for hand held terminals.

[3]   Conditional Access

The DVB did not standardise the entire area associated with paying for programming material. Hence the Subscriber Management System (SMS) where all customer data is stored, nor the Subscriber Authorisation Service (SAS), that encrypts and delivers the code words to allow descrambling, were standardised. In fact the DVB only standardised the 'Common Scrambling Algorithm' described below:

Common scrambling algorithm [ETR 289]

This tool ensures secure scrambling of the transport stream. Due to security issues the algorithm is only made available to those that need to know via a *custodian*, by way of a process described in [A 011].

Common Interface (CI) for Conditional Access [EN 50 221]

This specification defines an interface to the STB that allows the box to accept any CA system from different providers. The guidelines for use of the CI are given in [R206 001]. The CI interface is a powerful input / output device and so can also be used for other purposes. Such extensions are described in [TS 101 699].

DVB-GA 2 (94) 9 rev 1

Some STB's are designed with a single CA system built-in. A contract can however allow for that STB to receive programming from another provider. A basic contract that allows simulcrypt is given under this specification.

DVB-SIM [TS 101 197-1] and [TS 103 197]

The specification gives the head-end architecture of a simulcrypt system and how to synchronise different CA systems. Head-end control and management architectures are given in the second of the specifications given above.

[4]   Interactive Services

To allow for interactive services there are two distinct specifications; network dependent and network independent:

DVB-NIP [ETS 300 802]

This network independent specification defines an interaction enabling protocol stack. An important part of which came from the Digital Storage Media Command Control (DSM-CC) protocols created by MPEG [ISO/IEC 13818 – 6]. The [TR 101 194] document was then published as a guideline for use of this complicated software stack.

DVB-RCT [ETS 300 801]

This specification describes ways how to PSTN and ISDN networks as physical networks form interaction.

DVB-RCDECT [EN 301 193]

This describes the use of cordless telephone systems adhering to the DECT standard to enable return and interaction channels.

DVB-RCGSM [EN 301 195]

This describes the use of mobile telephone systems adhering to the GSM standard to enable return and interaction channels. Particularly useful for mobile use of DVB systems.

DVB-RCC [EN 300 800]

This specification, defined jointly by the DVB and DAVIC describes the use of CATV networks for bi-directional data communications. Guidelines for its use are given in [TR 101 196].

DVB-RCL [EN 301 199]

This is a derivative of the DVB-RCC specification and explains interactive services based on local multipoint distribution services.

DVB-RCCS [TR 101 201]

This gives guidelines or enabling the interactive channel based on satellite and coaxial sections distribution systems. This can also be used in some SMATV distribution systems.

DVB-RCS [TM 2267]

This describes an interaction channel via satellite.

[5]   Miscellaneous

DVB-DATA [EN 301 210]

This specification defines the five application areas of data *piping, data streaming, multiprotocol encapsulation, a data carousel,* and an *object carousel.* This for the distribution of securely protected data including the possibility of repeat transmissions at the same or irregular time periods. This also has a guideline document to describe its implementation: [TR 101 202].

DVB-DSNG [EN301 222]

This describes the transmission channel from a Digital Satellite News Gathering (D-SNG) unit to the central facility. [EN 301 210] describes the framing structure, channel coding and modulation of the channel.

PDH, SDH, ATM networks [ETS 300 813/4]

The digital telecommunications networks play an important role in connecting various facilities, for example a broadcaster's play-out centre and the satellite uplink station in some other city. The interface to these networks has therefore been defined in [ETS 300 813] for PDH, in [ETS 300 814] for SDH, and TR 100 815 for ATM.

DVB-M [ETR 290]

Testing of complete systems from broadcast to reception require certain test and measurement guide lines. These are given in this specification. Certain specific test and measurement signals can be incorporated into the transport stream, via a special PID value. [TR 101 291] explains its use.

STB interfaces [EN 50 201]

The DVB does not define the specification of the STB, however certain interfaces are defined.

Head end interfaces [EN 50 083 – 9]

This specifies the various interfaces at the cable head end, satellite uplink stations and other professional installations.

HAN [TS 101 224]

This specification describes the Home Access Network with active termination. Data being transmitted to the home or office. Also described in [TS 101 225] are the In-Home Digital Network (IHDN) and the Home Local Network (HLN). [TR 101 226] gives implementation guidelines for the IHDN.

Storage media [ETR 154] and [EN 300 468]

For STB's to interoperate with future types of storage media certain conditions must be met by the DVB streams. Eg the maximum bit rate for program transmission. Also the recording capabilities of the storage media. This specification details these conditions.

References

Books

[1]   H. Benoit: *Digital Television MPEG-1, MPEG2 and principles of the DVB system*. (Wiley).

[2]   David Peterson: *Audio, video and data telecommunications*. (McGraw-Hill)

[3]   Ferrel G. Stremler: *Introduction to Communications systems, 2nd edn*. (Addison-Wesley)

[4]   H. Pain: *The physics of vibrations and waves, 2nd edn*. (Wiley).

[5]   John B. Anderson: *Modern electrical communications*. (Prentice Hall).

IEEE publications

[1]   B. Hirosaki: *An orthogonal multiplexed QAM system using the discrete Fourier transform*. (Trans. Comm., Vol. 29, pp 982-989, July 1981).

[2]   Kalet: *The multitone channel*. (Trans. Comm., vol. 37, pp 119 - 124, Feb 1989).

[3]   W. Zou, Y. Wu: *COFDM : An overview*. (Trans. On broadcasting, vol. 41, no. 1, March 1995).

[4]   S. Darlington: On digital single-sideband modulators. (Trans. on cct. theory, vol. 17, no. 3, August 1970).

[5]   S. Weinstein, P. Ebert: *Data transmission by frequency division multiplexing using the discrete Fourier transform*. (Trans. comm. tech., vol. 19, pp 628 - 634, Oct 1971).

[6]   B. Slatzberg: *Performance of an efficient parallel data transmission system*. (Trans. comm. tech., vol. 15, pp 805 - 811, Dec 1967)

[7]   B. Le Floch, M. Alard, C. Berrou: *Coded orthogonal frequency division multiplex* (Proceedings, vol. 83, no. 6, pp 982 - 996, June 1995)

[8]   H. Sari, G. Karam, I. Jeanclaude: *Transmission techniques for digital terrestrial TV broadcasting.* (IEEE communications magazine, Feb 1995).

[9]   C. Del Tosso et al.: *0.5 μm CMOS circuits for demodulation and decoding of an OFDM based digital TV signal conforming to the European DVB-T standard.* (Jour. solid state ccts., vol 33, no. 11, November 1998)

Standards, specifications and other documents

[1]   EN 300 744 v1.1.2 DVB *framing structure, channel coding and modulation for digital terrestrial television*

[2]   ISO/IEC 13818-1: *Generic coding of moving pictures and associated audio: systems*

[3]   ETS 300 468: *DVB specification for service information in DVB systems*

[4]   R. W. Chang: *Orthogonal frequency multiplex data transmission system; patent number 3488445* (United States Patent Office 6 Jan 1970).

[5]   L. Møller: *Digital terrestrial television.* (EBU review, winter 1995)

[6]   DVBIRD: *Edited specification of the first generation receiver.*

[7]   Various internal STMicroelectronics documents.

[8]   Dr. Yiyan Wu: *Performance comparison of ATSC 8-VSB and DVB-T COFDM transmission systems for digital television terrestrial broadcasting.*

[9]   Australian department of communications and the arts: *Results summary of Australian 7 MHz laboratory tests of DVB-T and ATSC DTTB modulation systems.*

[10] Advanced television systems committee (ATSC): *ATSC digital television standard (Doc. A/53).*

[11] ITU-R draft new recommendation (TG 11/3-XXD) 1996 *Error correction data framing, modulation and emission methods for DTB.*

[12] DVB document A081 (DVB-H)

Glossary of abbreviations, terms and expressions

AC3	Multichannel digital audio standard developed by Dolby ®.
ACI	Adjacent channel interference. Interference from an adjacent sub carrier.
ADC	Analogue to digital converter.
AGC	Automatic gain control. The gain of a system is automatically adjusted to compensate for the variations present in an incoming signal.
API	Application programming interface. A software layer that presents a clean well defined set of functions to allow higher software layers to communicate and drive the layers lower down.
ASK	Amplitude shift keying. A modulation method that uses the amplitude of a signal to encode information.
ATM	Asynchronous transfer mode. Digital network system for the delivery of broadband-integrated services.
ATSC	Advanced television systems committee. A US organisation set up to set standards for the delivery of high definition television (HDTV).
AWGN	Additive white Gaussian noise.
BAT	Bouquet association table. This is a DVB table that lists the data services that can be grouped together for presentation to a viewer.
BE	Back end. A term relating to the back end electronics of a set top box. Ie from the transport stream input to the TV encoder output.
Bel	A unit of measurement of sound intensity and power levels in general.
BER	Bit error ratio. The ratio between the number of bits in error and the total number of bits transmitted.

BIOP	Broadcast inter-ORB protocol.
B-frame	Bi-directionally predicted frame. MPEG 2 frames that are coded with respect to I and P frames that precede and follow them.
BISS	Basic interoperable scrambling system. The DVB common scrambling algorithm for the scrambling of data in the transport stream for security purposes.
BPF	Band pass filter. A filter that only allows a certain band of frequency components to pass.
Byte-code	Java low level instructions that run on a Java virtual machine (VM).
CA	Conditional access. A means of allowing or not allowing programs to be decrypted at a receiver.
CAT	Conditional access table. A DVB table that gives the PID values of the packets containing the EMM's.
Cache	A small fast SRAM associated with a CPU used to buffer information from a slower external memory, thus making it more quickly available when needed.
CAT	Conditional access table. This is a DVB transport stream table that lists the PID values that contain the ECM messages.
CFT	Continuous Fourier transform. A method of converting continuous signals from the time domain into the frequency domain.
Circle surround ™	An SRS labs technology that allows a multi channel audio stream to be encoded as two channels for transmission (or storage), then reconstituted into the multi channels at the receiver.
CODEC	Coder / decoder. Generally a hardware that is able to perform coding and decoding together.
COFDM	Coded orthogonal frequency division multiplex. (OFDM with channel coding).

Convolution Coding	A method to add redundant information that can be used for error correction at the receiver.
CPE	Common phase error. A phase error common to a whole group of carriers, but uncorrelated from one OFDM symbol to another.
CPU	Central processing unit. The central part of a microprocessor that interprets the low level instructions.
CRC	Cyclic redundancy check. A byte or bytes added to the end of a longer word to allow for error checking and/or correction at a later stage.
CS™	Circle surround. SRS labs audio standard for transmission of multi channel audio through a two channel delivery system.
CW	Continuous waveform. Cosine or sine wave.
dB	Decibel, or 1/10th of a bel. A unit of measurement of sound intensity and power levels in general.
dBm	Decibel miliwatt. A unit of power to describe a signal strength (with respect to a signal of power 1mw).
dBµV	Decibel microvolt. A unit of power to describe a signal strength (with respect to a signal of level 1µV).
DAC	Digital to analogue converter.
DCT	Discrete cosine transform. A particular case of a Fourier transform applied to discrete sampled signals. The result is a series of sine and cosine terms with varying amplitude coefficients.
DEMUX	Demultiplexing. The action of splitting data up depending on its content type, and sending to different areas for processing
DENC	Digital encoder. The electronics responsible for producing the PAL, NTSC or SCAM signals needed as input to analogue TVs.

DFT	Discrete Fourier transform. A method of converting a sampled signal from the time domain to the frequency domain.
DMA	Direct memory access. A circuit block that accesses memory directly to move the contents to another location, independently of the CPU.
Dolby ®	A US organisation with a particular audio compression algorithm used widely in the field of consumer audio / video products.
Doppler shift	The change of frequency associated with the movement of the transmitter with respect to the receiver.
DPSK	Differential phase shift keying. A modulation technique.
DSB-SC AM	Double sideband suppressed carrier amplitude modulation.
DSM-CC	Digital Storage Media Command Control. A set of protocols to define the user to user transmission structure. Enables the object carousel software protocol. (Specified in ISO/IEC 13818-6[26]).
DSS	Digital Satellite System. First digital satellite broadcasting system developed by Hughes in the US.
DTG	Digital television group. A British organisation setup to make DTV a reality in the UK.
DTS ®	Digital theatre sound. Technology developed by Digital theatre systems Inc.
dTTb	Digital terrestrial television broadcast. A European group set up to define and prove a digital TV transmission standard.
DTTV	Digital terrestrial television.
DVBird	Digital video broadcast integrated receiver decoder. A project to construct the first
COFDM	based digital TV receiver.
DVB	Digital video broadcast

DVB-C	Digital video broadcast – cable
DVB-S	Digital video broadcast - satellite
DVB-T	Digital video broadcast – terrestrial
DVB-H	Digital video broadcast - handheld
DVD	Digital versatile (or video) disc
ECM	Entitlement control message. This is a conditional access message defined by the DVB standard.
EIT	Event information table. This is a DVB table that gives information regarding an event.
EM	Electro magnetic. The nature of waves such as light, radio, X-ray, and micro wave etc.
EMM	Entitlement management message. A conditional access message defined by the DVB standard.
EPG	Electronic program guide. The TV guide allowing the viewer to easily select programs of interest.
ES	Elementary stream. The raw digital programming data streams.
FCC	Federal communications commission.
FDM	Frequency division multiplexing. A method of splitting data up into chunks and transmitting these with different carrier frequencies.
FE	Front end. A term relating to the electronics of the front end of a set top box. Ie from the input tuner to the output transport stream prior to input to the demux process.
FEC	Forward error correction. The addition of redundant data to a bit stream prior to transmission, allowing for errors due to transmission to be corrected at the receiver.
FFT	Fast Fourier transform. An algorithm for computing the discrete Fourier transform (DFT) more quickly.

FIFO	First in first out. Electronic component that act as a buffer. When reading, the first data to be written in is read out.
FLASH	A non volatile memory that can be programmed in situation on the PCB generally with no special voltage levels needed. Erasing and reprogramming is generally achieved in blocks.
FSK	Frequency shift keying. A method of modulation that uses changes is frequency to represent different bit patterns.
GOP	Group of pictures. A term used in MPEG compression to describe the pictures within two intra frame (I- frame) pictures.
Grand Alliance	An alliance formed between AT&T, David Sarnoff research centre, GI corp., MIT, Philips electronics, Thomson consumer, and Zenix electronics corp. To advance the adoption of the 8-VSB system for digital TV transmission.
Hamming distance	Binary words are said to be close to each other if the hamming distance between them is small. This distance is a measure of how many symbols in the word are different. Mathematically: $h = w (S_a \oplus S_b)$. Where S_a are the symbols of cod a and S_b are the symbols of word b. w simply represents an operator that returns the number of '1's in the resulting word. $\oplus$ signifies modulo two addition.
HD	High definition.
HDTV	High definition television.
HPF	High pass filter. A filter that only allows high frequency components to pass.
HRTF	Head related transfer function. An SRS labs term related to the human perception of audio.
HTML	Hypertext mark-up language.
HTTP	Hypertext transfer protocol.

Huffmann Coding	Also known as variable length coding (VLC). Infrequently occurring numbers are assigned a small length word, and more frequently occurring numbers are assigned a longer word. This is a method of compression
ICFT	Inverse continuous Fourier transform. A method of converting continuous signals from the frequency domain to the time domain.
IDFT	Inverse discrete Fourier transform. A method of converting a sampled signal from the frequency domain to the time domain.
IDTV STB	Integrated digital television. A TV with the processing integrated inside the TV.
IF	Intermediate frequency. The first (or any) stage of frequency reduction inside a tuner.
IFFT	Inverse fast Fourier transform. A fast method of converting continuous signals from the frequency domain to the time domain.
I-frame	Intra-frame. MPEG 2 frames coded with respect to no other frames other than themselves.
I/O	Input / output.
IP	Intellectual property, or Internet protocol.
IRD	Integrated receiver decoder.
ISDN	Integrated services digital network.
ISI	Inter symbol interference. Interference within a particular symbol.
Java	A general purpose object oriented concurrent programming language developed by Sun Microsystems Inc.
Java bean	A Java object that obeys certain rules
LED	Light emitting diode
Lossy	A term used to describe a system of compression that loses data. Ie if the compression is performed a number of times, then decompression is performed a number of times, the data will not be as the original.

LPF	Low pass filter. A filter that only allows low frequency components to pass.
MCM	Multi carrier modulation.
MFN	Multi frequency network. A broadcasting method that utilises more than one frequency to cover a large area.
MHEG	Multimedia hypermedia experts group. Standard set to support the distribution of interactive multimedia applications in server / client architectures.
MHP	Multimedia Home Platform. DVB specific STB middleware for interactive services.
Middleware	A software layer used inside a STB. A layer on which the application programs can run.
Modulo 2 Addition	Binary addition that takes the form:

$$0 \oplus 0 = 0$$
$$1 \oplus 0 = 1$$
$$0 \oplus 1 = 1$$
$$1 \oplus 1 = 0$$

The $\oplus$ symbol signifies the modulo 2 addition.

MPE	Multiprotocol Encapsulation
MPEG	Moving pictures expert group. A group set up to set standards in the area of video and audio compression.
MP@ML	Main profile at main level. One of the MPEG compression standard options.
NIT	Network information table. A DVB table that gives information of a network made up of more than one transport stream.
NRZ	Non return to zero. A method of describing a bit pattern. For example a high level is used to represent a 1, but remains high if another 1 is to be represented, but goes to a lower level to represent a 0. Hence no return to a zero level after a particular bit has been represented.

NTSC	A TV transmission standard used in the US and other countries.
NVOD	Near video on demand. The transmission of the same program a number of times, but spaced out in time. Thus allowing a viewer to see the program whenever he/she wants within limitations.
OFDM	Orthogonal frequency division multiplex. A method of data transmission.
Orthogo- nality	A term used to describe the complete independence of two or more signals or functions.
OSD	On screen display. The graphics display of menus and images on a TV screen.
PAL	A TV transmission standard used in Europe and other countries.
PAS	Program association section. These are DVB sections that make up the PSI tables.
PAT	Program association table. This Is a DVB transport stream table that is transmitted with unique PID value 0, and contains a list of which program numbers are associated with which PID values.
PCR	Program clock reference. A value that is sent with a digital program stream as a means of achieving receiver synchronisation at the program level.
PES	Packetised elementary stream. The raw digital programming data streams inside a transport packet.

P-frame	Predicted frame. MPEG 2 frames that are coded with respect to I and P frames that precede them.
Phase noise	The phase of modulated signals can vary with respect to the theoretical, this is know as phase noise.
PID	Program identifier. A unique code in the DVB transport stream defining a particular elementary stream.
Pilot	A pilot carrier is a carrier that contains information either by virtue of its position or in its data content on the channel such that correction processes can be implemented at a receiver.
PDH	Plesiochronous digital hierarchy. Digital telecommunications network.
PLL	Phase locked loop. A method of producing a constant frequency.
PMS	Program map section. These are DVB sections that are used to construct the PMT.
PMT	Program map table. A particular program in a DVB transport stream will have a number of different elementary steams associated with it and given unique PID values. The program map table lists which programs are associated with which PID's.
Polynomial	A mathematical function made up of many terms.
PPV	Pay per view.
PRBS	Pseudo random binary sequence.
Pro-Logic surround	A Dolby® labs audio standard to give a sound effect with only two speakers.
PSTN	Public switched telephone network.
PSK	Phase shift keying. A modulation method that uses the phase of a signal to encode information.

PTI	Programmable transport interface. An interface toaccept and perform some preliminary processing on the MPEG 2 transport steam.
Puncturing	A means of reducing the amount of redundant data.
PVR	Personal Video Recorder (Generally a STB with a hard disc drive to the storage of program material
PWM	Pulse width modulation. The encoding of information onto a signal by varying the mark/space (high/low level) of a signal.
QAM	Quadrature amplitude modulation. A modulation method that uses both phase and amplitude to encode information.
QEF	Quasi error free. The term used to describe a communications channel that limits errors to around 1 in every 10^{10} bits transmitted.
QPSK	Quadrature phase shift keying. A modulation method that uses 4 different phases to encode information.
Quantisation	The act of ascribing code words to DCT coefficient values as part of the MPEG 2 compression algorithm.
RGB	Red, green, blue. Colour signals used as input to a TV.
RLE	Run length encoding. Strings of 1's or 0's can be encoded as the number of 1's or 0's there are in a row. Transmitting (or storing) this number will result in less data being sent (or stored) than sending (or storing) the rows of 1's or 0's themselves.
RS	Reed-Solomon. The scientists associated with a particular error correction technique used widely in the area of digital transmission.
RST	Running status table. This is a DVB table that allows updates of timing status of one or more events.

SAS	Subscriber Authorisation System. The system that encrypts and delivers the code words that allow a receiver to descramble programming material.
S/C	Signal to carrier ratio.
SD	Standard definition.
SECAM	A TV transmission standard used in France.
SFN	Single frequency network. A method of broadcasting using only one frequency to cover a wide area, generally with many transmitters.
SDH	Synchronous digital hierarchy. Digital telecommunications network.
SDRAM	Synchronous dynamic random access memory. This is a fast DRAM memory device. All the control strobes being generated internally to the device from a single fast external clock.
SDT	Service description table. This is a DVB table that lists various parameters associated with service data.
SDTV	Standard definition television.
SI	Service information. Information transmitted in a DVB transport stream.
SMATV	Satellite Master Antenna Television. A media distribution system.
SMS	Subscriber Management System. The system that stores all the subscriber personal information.
S/N	Signal to noise ratio.
SNR	Signal to noise ratio.
S/PDIF	Serial / Parallel Differential. Audio output. Normally coaxial and optical for the output of compressed or non compressed digital audio data to an external amplifier / decoder.
SRAM	Static random access memory. Fast memory used in many electronic systems.
SRS®-3D	SRS labs audio standard to give a 3D sound effect from a stereo input source and two speakers only as output.

SSB-AM	Single sideband amplitude modulation.
ST	Stuffing table. A table that is used to invalidate section data.
STB	Set top box. Any system that receives and decodes digital program streams at the home.
TCP/IP	Transmission control protocol/Internet protocol.
TOT	Time offset table. This is a DVB table that gives actual time information.
TOV	Threshold of visibility.
TPS	Transmission parameter signalling. COFDM system carriers that carry system related information.
Trellis Coding	Used in ATSC 8-VSB coding. A method to add redundant information that can be used for error correction at the receiver.
Tru-Surround™	An SRS labs audio standard for multi channel audio on two speakers.
UART	Universal asynchronous receiver/transmitter. A system allowing digital data to flow in and out.
UDP	User datagram protocol. A protocol to realise broadcast and interaction channels (UDP/IP).
UHF	Ultra high frequency (300 MHz to 3 GHz).
VCXO	Voltage controlled crystal oscillator. A device that allows a particular frequency to be produced by applying a particular voltage.
Vector	A representation of a force or a signal which has magnitude and direction.
VHF	Very high frequency (30 MHz to 300 MHz)
Viterbi	A scientist involved with convolution code theory.
Viterbi algorithm	Ensures the path taken through a convolution decoder is the minimum error path, thus allowing for errors to be identified and corrected. (Maximum likelihood decoding).

VLC	Variable length coding. Infrequently occurring numbers are assigned a small length word, and more frequently occurring numbers are assigned a longer work. This is a method of compression. Also termed Huffmann coding.
VL-RISC	Variable length reduced instruction set computer. A CPU that benefits from the smaller instructions of a RISC architecture, but can construct more complicated (8, 16, 24 and 32 bit) instructions when required.
VM	Virtual machine. A software program that mimics a CPU by allowing applications programs to run on top of it. MHP implements the Java VM.
VOD	Video on demand.
VSB	Vestigial sideband modulation. A form of amplitude modulation used for the transmission of data.
YUV	Luminance, blue chrominance and red chrominance or colour difference. Sometimes also known as Y, C_b, C_r. These are the pixel information signals transmitted to black and white and colour TVs alike.
XOR	Exclusive or. A logical operation that gives a high output only if one or the other of two inputs is high, but not if neither nor both are high

INDEX

8

8-VSB, i, viii, ix, 1, 3, 51, 148, 149, 150, 152, 155, 162, 165, 171, 172, 173, 174, 175, 177, 179, 180, 181, 182, 186, 233, 234, 250, 256, 263

A

B

C

Diagnostic controller, 115
Diffusion, 124
Digital encoder, 114
Digital FDM, 93
Diversity, 204
DMA, 111, 254
Dolby®, ix, 187, 192, 193, 194, 195, 196, 242, 251, 260
Doping, 124, 125
Doppler, ix, 69, 70, 177, 178, 206, 207, 254
DPSK, 73, 254
DSB-SC, 237, 254
DSM-CC, 132, 137, 139, 140, 245, 254
DTS, 17, 19, 192, 254
DVB, i, vii, ix, 1, 3, 26, 46, 47, 48, 50, 51, 52, 56, 57, 64, 73, 74, 76, 77,
    78, 81, 89, 107, 126, 172, 173, 174, 175, 176, 178, 179, 180, 181, 182,
    183, 212, 217, 249, 250, 251, 252, 254, 255, 258, 259, 260, 261, 262,
    263
DVB-C, 3, 107, 139, 197, 243, 255
DVB-H, x, 68, 143, 204, 205, 207, 243, 250, 255
DVBird, 4, 254
DVB-S, 3, 46, 47, 77, 107, 139, 197, 242, 243, 244, 255
DVB-T, i, vii, ix, 1, 3, 46, 47, 48, 50, 51, 52, 56, 57, 64, 74, 76, 77, 78, 81,
    89, 107, 126, 172, 173, 174, 175, 176, 178, 179, 180, 182, 183, 212,
    217, 250, 255
DVD, 187, 193, 200, 201, 202, 203, 255

E

Echo, 64, 69, 175, 176, 177, 178, 179, 181, 204
Echo correction, 179
ECM, 26, 28, 119, 120, 121, 252, 255
EIT, 10, 19, 28, 255
Electronic program guide, 1, 27, 109, 112, 129, 197
Elementary stream, 7, 10, 13, 19, 259
EMI, 114, 198
Encryption, 115, 116, 117, 121
Energy, 47, 54, 77, 78, 106, 151, 154, 175
Energy dispersal, 106, 151
EPG, 27, 28, 109, 112, 126, 129, 130, 132, 197, 255
Error, 6, 8, 9, 36, 43, 46, 47, 50, 51, 54, 55, 56, 70, 71, 78, 100, 102, 105,
    106, 112, 126, 149, 150, 151, 153, 154, 156, 158, 159, 166, 168, 174,
    175, 200, 206, 251, 253, 255, 261, 263
Error correction, 6, 43, 46, 55, 56, 105, 126, 149, 150, 151, 153, 168, 175,
    200, 206, 253, 255, 261
ES, 7, 19, 255
Essence, 40, 195

Event information table, 19

Multi carrier modulation, 2
Multi frequency network, 179
Multicrypt, 121

N

Near video on demand, 259
Network information table, 24, 108
Network PID, 24
NIT, 10, 27, 108, 258
NRZ, 213, 233, 258
NTSC, 76, 114, 148, 182, 191, 253, 259
NVOD, 259

O

OFDM, vi, vii, 2, 3, 54, 58, 59, 60, 61, 66, 73, 79, 80, 81, 84, 90, 91, 99,
    101, 102, 103, 250, 252, 253, 259
OFDM demultiplexing, vii, 101
OFDM patent, 2
Offset, 10, 17, 22, 29, 56, 99, 101, 158, 178, 263
OPCR, 14, 16
Operating system, 109, 128, 129
Optional field, 14, 18
Orthogonality, vi, x, 91, 232
OSD, 187, 191, 259
Outer coding, 47, 48, 78, 152, 175
Outer decoding, 78, 105
Oxidation, 123

P

PAL, 76, 114, 191, 253, 259
PAS, 20, 259
PAT, 10, 20, 23, 24, 26, 27, 29, 108, 117, 259
PCM/CIA, 116, 121
PCR, 7, 14, 15, 16, 28, 106, 112, 259
PDH, 247, 260
PES, 8, 9, 18, 26, 117, 198, 200, 259
P-frame, 260
Phase mappings, 79
Phase noise, 71, 99, 111
Phasor, 79, 215
Photo resist, 124
PID, 10, 12, 13, 16, 20, 24, 26, 27, 28, 29, 110, 114, 117, 118, 247, 252,
    259, 260
Pilots, 70, 71, 99, 102, 104

Q

R

RST, 10, 29, 261
Running status table, 261

S

S/PDIF, 201, 202, 262
Scattered pilots, 99, 102, 104
Scrambling, 11, 18, 44, 45, 106, 115, 118, 244, 252
SDH, 247, 262
SDRAM, 110, 143, 144, 187, 262
SDT, 10, 28, 262
SECAM, 191, 262
Section filter, 110, 137
Service information, 10, 19, 250
Set top box, 1, 4, 93, 109, 115, 123, 126, 193, 197, 251, 255
SFN, 55, 57, 61, 62, 64, 66, 74, 173, 179, 207, 243, 262
Shannon, 43
SI, v, 10, 19, 26, 28, 206, 242, 262
Sidebands, 236, 237
Silicon, 123, 124, 125, 127, 129, 141, 198, 200
Simulcrypt, 121, 244
Single frequency network, 55, 57, 61, 62, 69, 179, 243
Smart card, 113, 116, 118, 119, 121, 128
SMATV, 243, 246, 262
SNR, 32, 104, 262
Software architecture, vii, viii, 128
Spectrum, vi, viii, 54, 55, 98, 148, 218, 225, 234, 236, 237, 240
Splice countdown, 14, 16
Splicing point, 14, 16
SRAM, 114, 252, 262
SRS®, ix, 193, 196, 252, 253, 256, 262, 263
SSB-AM, 237, 263
ST, 10, 29, 131, 242, 263
ST20, 109, 129
STB, vii, ix, 4, 29, 94, 110, 111, 116, 117, 118, 121, 122, 128, 129, 130,
    131, 132, 140, 141, 142, 143, 145, 193, 197, 198, 200, 201, 242, 244,
    247, 248, 257, 258, 261, 263
STMicroelectronics, 109, 129, 130, 198, 250
Stuffing, 10, 13, 22, 44
Stuffing table, 10
Super frame, 60, 68
Surround sound, 193
Symbol, 52, 58, 59, 60, 61, 63, 65, 66, 67, 69, 71, 73, 77, 79, 82, 90, 91,
    99, 100, 102, 103, 104, 159, 162, 165, 170, 171, 217, 253, 257, 258
System on a chip, 126

T

U

V

www.ingramcontent.com/pod-product-compliance
Lightning Source LLC
Chambersburg PA
CBHW052012230326
41598CB00078B/2807